EVOLUTION NOW

D0973411

must-
variatn.
a heredity

EVOLUTION NOW

A CENTURY AFTER DARWIN

EDITED BY

John Maynard Smith

in association with

nature

W. H. Freeman and Company
San Francisco

Library of Congress Cataloguing and Publication Data
Main entry under title:

Evolution now.

1. Evolution – Addresses, essays, lectures.
2. Life – Origin – Addresses, essays, lectures.
1. Maynard Smith, John, 1920-
QH 366·2·E857   575   82-2470
ISBN 0-7167-1426-4   AACR2
ISBN 0-7167-1427-2   (pbk).

© *Nature* 1982

Published in the United States in 1982 by W. H. Freeman and
Company, 660 Market Street, San Francisco, California 94104.
First published in Great Britain in 1982 by *Nature* in association
with the Macmillan Press Ltd.

Reprinted 1983

Printed and bound in Great Britain.

CONTENTS

INTRODUCTION

LUDWIG Boltzmann once wrote that the nineteenth century would be remembered as the century of Darwin. One hundred years after Darwin's death this judgement still seems perceptive. No other writer had such a profound effect on the way we see ourselves, and no other brought about so great an extension in the range of subjects which we regard as explicable by scientific theory. Here, however, I shall confine myself to his contribution to evolutionary biology, and shall forget that he was the founder of ecology and ethology, and made significant contributions to geology and psychology. It was, after all, his formulation of the theory of evolution by natural selection that was decisive.

In recent years there have been claims—in the daily press, on television, and by retired cosmologists—that Darwin may have got it wrong. Some excuse can be found in the fact that Darwin has indeed been criticised by scientists working in a variety of fields—for example palaeontology, taxonomy and embryology. At least one group of scientists have claimed that a new evolutionary paradigm is on the way. The most controversial of these issues are debated in this book. However, to see Darwinism as being under serious threat would, I think, be a false perception. The error arises because Darwin's theory is so central to modern biology that any new idea may first be seen (as Mendelian genetics was seen) as being in conflict with Darwinism.

This volume presents some current controversies and recent advances in evolutionary biology, by reprinting papers published in the last few years. But let me, as a background, first give a brief history of evolutionary ideas since Darwin. In the *Origin of Species*, Darwin aimed to establish two things. First, he argued that evolution had in fact happened (that is, that all existing organisms are descended from one or a few simple ancestral forms), and, second, that the main cause of evolutionary change was the natural selection of variations that were in their origin non-adaptive. The main weaknesses of his position were that he had no adequate theory of genetics, and that he could give no satisfactory account of the origin of the variations on which selection would later act. In genetics, Darwin was a Lamarckist. That is, he thought that if organisms acquired characteristics by use and disuse during their lifetimes, this would influence the nature of their offspring. In thinking this, he was sharing an opinion held by almost all his contemporaries.

The first major advance after Darwin was made by August Weismann, who argued for the independence of the 'germ line' (the cell lineage leading from the fertilized egg to the germ cells, egg and sperm, which form the starting point of the next generation) from the 'soma' (the cell lineage from fertilized egg to adult body). As a boy, I acquired, from reading the preface to Shaw's *Back to Methuselah*, a picture of Weismann as a cruel and ignorant German pedant who cut the tails off mice to see if their offspring had tails. What a ridiculous experiment! Since the mice did not actively suppress their tails as an adaptation to their environment, no Lamarckist would expect the loss to be inherited. Much later, I discovered that Weismann was not as I had imagined him. His experiment on mice was performed only because, when he first put forward his theory, he was met with the objection that, (as was, it was claimed, well known) if a dog's tail is docked, its children are often tailless—an early use of what J.B.S. Haldane once called Aunt Jobisca's theorem, 'It's a fact the whole world knows'.

Much more interesting are Weismann's reasons for proposing his theory, and its implications for Darwinism. At first sight, his reasons were poor. It is often not the case (for example, in higher plants) that the germ line is a lineage distinct from the soma. Even when it is distinct, the material and energy it needs are supplied by the soma. In Weismann's day, the experimental evidence for the 'non-inheritance of acquired characters' was weak. Why, then, did he believe it? I think that the clue lies in his remark that, if one were to come across a case of the inheritance of an acquired character, it would be as if one were to send a telegram to China and it arrived translated into Chinese. This is the first use known to me of the information analogy in heredity. Weismann did not accept the inheritance of acquired characters because he could not conceive of a mechanism of 'reverse translation', whereby the hypertrophied muscles of the blacksmith could be translated into genes (he called them 'ids') which could, in the next generation, cause the growth of large muscles.

If Weismann was right, this greatly strengthened Darwin's theory. Natural selection, instead of being just one of the possible processes leading to evolutionary adaptation, becomes the only process (at least, until the evolution of organisms sufficiently intelligent to learn from their parents).

The next important step was the rediscovery of Mendel's laws at the start of this century, and the formulation of the chromosome theory of heredity. The first impact of Mendelism on evolutionary biology was distinctly odd. The early Mendelians saw themselves as anti-Darwinians; Darwin's banner was held aloft by the biometric school, who concentrated on measuring the correlations between relatives, and who regarded genes as metaphysical entities. The Mendelians saw the 'mutations' they studied as each being the potential starting point of new species, and the continuous variation studied by the biometricians as evolutionarily irrelevant; the biometricians saw mutations as pathological deviants doomed to early elimination by selection, and continuous variation as the stuff of evolution. The argument, led by Bateson on the one side, and Pearson on the other, foreshadowed the current debate between punctuationists and gradualists, discussed below. It is part of the larger debate between those who see the world as continuous and those who think it proceeds in jerks.

The debate, at least in the form in which it then presented itself, was largely settled by the work of the population geneticists, Fisher, Haldane and Wright. Two points were made clear. First, the continuous variation studied by the biometricians could be explained by alternative alleles at many loci, each by itself having a small effect on the phenotype. Second, even rather small differences in fitness between genotypes are sufficient to determine the direction of evolutionary change, despite mutation being mainly in an opposite direction.

The work of the population geneticists prepared the way for the 'modern synthesis' of evolutionary biology, developed in the period 1930–1950 by a group including Dobzhansky, Ford, Julian Huxley, Mayr, Muller, Rensch, Simpson and Stebbins. It is hard in a few sentences to describe what these men did. In effect, they showed that the 'neo-Darwinian' mechanism—natural selection in Mendelian populations—was sufficient to explain the evolutionary process as it could be observed in nature. Dobzhansky, Ford and others measured genetic variability and natural selection in wild populations. Mayr and Rensch (for animals) and Stebbins (for plants) studied geographical variation within and between species, and discussed how new species might arise. Simpson argued that the fossil record could best be understood in

Darwinian terms. Most research in evolutionary biology since that time has been carried out in the framework of the modern synthesis. Particular efforts have been made in areas which, at least at first sight, seem to be difficult to explain in terms of natural selection, for example, the evolution of social behaviour and of sex and breeding systems.

Since 1950, developments in molecular biology have had a growing influence on the theory of evolution. The 'central dogma' of molecular biology, according to which information can pass from nucleic acid to protein, but not from protein to nucleic acid, provides a molecular explanation for Weismann's principle, thus leaving natural selection as the only agent of adaptation. Important as this is, however, two additional points should be made. First, even if 'reverse translation' of amino acid sequences into base sequences were possible, this would not provide a general mechanism for Lamarckian inheritance, because most developmental adaptations do not involve the production of new protein sequences. Second, there are good reasons why, even if living organisms have arisen independently many times in the universe, Lamarckian processes should play a minor role in their evolution. Most 'acquired characters' are non-adaptive—they are the results of age, injury and disease. Therefore, a hereditary mechanism which transmitted such characters to offspring would work against the evolution of adaptation. Hence the one-way flow of information from nucleic acid to protein may have been a necessary feature of an hereditary mechanism able to support evolution. In physics, the second law of thermodynamics asserts that entropy will increase in a closed physical system. In biology, Weismann's principle, together with the principle of natural selection, makes possible the maintenance, and even the increase, of information in open biological systems.

Molecular biology has had an impact on evolutionary theory in other ways. Protein electrophoresis has provided a way of measuring the genetic variability of populations; its main value may be in enabling us to discover more about the breeding structure of populations. Sequence data on proteins and nucleic acids can be used to work out the phylogenetic relationships of existing organisms. In this context, sequences have the advantage over morphological data in that they provide a means of estimating the number of genetic changes separating two forms. The information we are acquiring about how DNA is arranged in chromosomes may at last give us some insight into the evolutionary significance of chromosome structure. Two questions in particular are being asked. First, does the large-scale arrangement of genes on chromosomes have any significance for development, or is it merely a way of ensuring accurate gene segregation during cell division? Second, does all the DNA in the genome perform some useful function in the survival or reproduction of the organism, or is some part of it 'selfish' or 'parasitic'? The second question raises a set of problems which are logically similar to those which have been debated for some time by students of the evolution of social behaviour, that is, questions about the levels at which selection acts and the differences between what Dawkins has called 'replicators' and 'vehicles'.

There is, however, one area of molecular biology which seems to me to lag behind the rest. This is the study of the evolution of prokaryotes (organisms such as bacteria lacking a proper cell nucleus). The modern synthesis of the 1940s was concerned with eukaryotes (organisms with a cell nucleus, usually sexual and diploid). Its essential achievement was to bring together two previously separate disciplines—the chromosome theory of heredity and the study of natural populations. The same synthesis is now required for the prokaryotes. There is an abundant knowledge of their genetics, but as yet no adequate synthesis of that knowledge with a study of the natural history of bacteria. For example, we have little idea of the significance of conjugation for bacterial populations; it is as if we had no idea of the significance of sexual reproduction for populations of birds or insects. Population thinking has been well developed for fully half a century, but has yet to be adopted by microbiology.

The papers printed below have been grouped under six topics, each with a brief introduction aimed at doing three things. First, I explain how the topic is related to Darwin's ideas. Second, I have tried to help non-specialists to find their way through papers which are sometimes rather technical. Some of the critical technical terms are defined in the introductory passages, but it is inevitable that some words will be unfamiliar to some readers. Nevertheless, I would urge readers to press on regardless; it should usually be possible to grasp the gist of the argument. Finally, I have allowed myself the indulgence of expressing my own opinion on some of the more controversial issues.

The selection of papers has inevitably been somewhat arbitrary. We have not aimed at reprinting just the most important papers published in recent years. Instead, we have concentrated on the more controversial fields at the expense of others in which equally valuable work is being done. Some topics have been omitted because of the difficulty of finding suitably brief and non-technical articles. I particularly regret that there is little or nothing on the evolution of breeding systems, on mechanisms of speciation, on the biology of islands or on continental drift—all subjects which would have fascinated Darwin. Finally, I must explain that I have not, in this paragraph, lapsed into the use of the Royal 'we': in selecting and editing the papers I have worked with Alun Anderson, Peter Newmark and Miranda Robertson of the editorial staff of *Nature*.

I am conscious that, by concentrating on controversial topics, we may have given unjustified prominence to ideas which will prove to be wrong. In mitigation, let me quote Darwin's remarks in *The Descent of Man*: 'False facts are highly injurious to the progress of science, for they often endure long; but false views, if supported by some evidence, do little harm, for everyone takes a salutory pleasure in proving their falseness; and when this is done, one path towards error is closed and the road to truth is often at the same time opened.'

CHAPTER ONE

THE ORIGIN OF LIFE

'THERE is a grandeur in this view of life, with its several powers, having been originally breathed by the creator into a few forms or into one: and that, whilst this planet has gone cycling on according to the fixed laws of gravity, from so simple a beginning endless forms most beautiful and most wonderful have been, and are being evolved.' When one considers the agnosticism of his autobiography and his notebooks, these words of Darwin's, the closing words of the *Origin of Species*, can only be seen as a sop to the Cerberus of orthodoxy. The origin of life was, in Darwin's day, inaccessible to scientific study—so why not credit it to the creator?

Today the problem of the origin of life, although far from being solved, is being actively studied, both experimentally and theoretically; we can no longer leave things to the breath of the creator. It turns out that although Darwin did not think seriously about the problem, his theory of evolution provides us with the only satisfactory definition of 'life', and hence with the only clear way of formulating the problem of its origins. Entities which have the properties of multiplication, variation and heredity are alive, and those which lack one or more of those properties are not. This definition is not arbitrary, because once entities arise which have these properties, populations of such entities will evolve by natural selection, and will acquire the other features of wholeness, self-maintenance, complexity, adaptation to the environment, and so on, which are associated with living organisms.

According to this definition, the RNA molecules which evolved in test tubes, as described in the paper by Eigen *et al.*, were alive: they had heredity, multiplication and variation, and consequently they evolved adaptations to the environment of the test tube. However, these experiments do not solve the problem of the origin of life, because it was necessary to supply a complex enzyme, Q_β replicase, which could not have been present in the primitive oceans: the molecules could only evolve in an environment which was informationally more complex than themselves. The experiments are nevertheless illuminating. RNA is a good candidate as a primitive replicating entity because, being singlestranded, it has a phenotype—it bends back on itself and forms hairpins, clover-leaves and so on—whereas double-stranded DNA is rather boring, as well as being difficult to replicate. What these experiments demonstrate is that nucleic acid molecules will evolve in fairly simple environments.

Where, then, do we stand? It seems that the problem of the origin of simple organic molecules—in particular, the amino acids, bases and sugars which are the components of proteins and nucleic acids—is solved in principle, if not in full detail. The problem of how such molecules formed polymers is somewhat harder; the main difficulty is that there would be a tendency to form irregular polymers, with varying types of chemical bond linking a range of component molecules, rather than the regular polymers found in existing organisms. Polynucleotides, once formed, might replicate by complementary base pairing as they do today, in the absence of 'informed' enzymes (that is, enzymes with specific amino acid sequences, adapted to ensure accurate replication, and themselves coded for by nucleic acids.) The snag is that they would replicate very slowly, which would not matter, and very inaccurately, which would. This leads to the catch-22 of the origin of life. One could not maintain a long polynucleotide sequence without informed enzymes, and one could not have informed enzymes without a long polynucleotide to code for them. It is this paradox which is the main topic of the article by Eigen *et al.* and of my own article discussing their work. A second major problem, which is touched on by Eigen *et al.*, is the origin of the genotype—phenotype distinction: that is, of Weismann's distinction between germ line (entities which are replicated, and whose information content is, at least potentially, immortal) and soma (an entity which is a translation of the information in the germ line, and which is mortal). The distinction is fundamental, because the kinds of objects which lend themselves to accurate replication are likely to be quite different from those which function effectively in ensuring survival and growth. If for no reason, this will be true because an entity which can be accurately replicated will, assuming template reproduction, be essentially one- or two-dimensional, whereas functionally efficient objects are likely to be three-dimensional. For the genotype-phenotype distinction to be established, however, requires that there be a process of development, during which genotypic information is translated into phenotypic structure.

The origin of genetic information

MANFRED EIGEN, WILLIAM GARDINER,
PETER SCHUSTER AND
RUTHILD WINKLER-OSWATITSCH

Laws governing natural selection of prebiotic molecules have been inferred and tested, making it possible to discover how early RNA genes interacted with proteins and how the genetic code developed

CHARLES Darwin saw in the diversity of species the principles of evolution that operated to generate the species: variation, competition and selection. Since Darwin's time an understanding of molecular biology and of the geophysics and geochemistry of the prebiotic earth has been gained that would have been unimaginable in the 19th century. Does that make it possible now to follow evolution back into the era before there were organisms?

A first answer is no. The prebiotic fossil record, as far as is known, decayed or was wiped clean by later generations of life. The intellectual fossils that remain—the genetic code, the genetic messages of present organisms and the known reaction pathways of biochemistry—convey information so fragmentary that one could never describe prebiotic evolution in as much detail as, say, the evolution of primates.

Fragmentary information, however, has never been a barrier to the discovery of laws of nature. Newton discovered the universal law of motion from observations of a few planets; Mendeleev discovered the structure of the periodic table in the chemistry of only a few elements; today physicists infer laws that describe the interactions of elementary particles from observations of small numbers of events. One does not need a detailed history of prebiotic conditions and events in order to discover the evolutionary laws that led to the first life on the earth. One must only hope there is enough fossil evidence left to guide one's thinking and demand enough predictive power of a theory to make it subject to experimental testing. In this sense the answer to the question raised above is yes: Definite statements can be made about the natural laws that governed the origin and prebiotic evolution of life.

In this article we describe what must be added to Darwin's ideas to describe evolution before organisms came into existence. First we shall show that his ideas do apply to evolution far below the level of organisms. To explain how the complexity of higher organisms and the variety of species came about, Darwin proposed that the more complex evolves from the less complex through natural selection. Why should this principle not apply as well to the complexity of large molecules? We shall give necessary and sufficient conditions for natural selection to proceed at the molecular level. The outcome of such selection is a 'regularity of events' that follows inevitably whenever certain conditions are fulfilled.

Whereas competition is the basis of natural selection among organisms, competition alone would not have worked in prebiotic times to select the fittest molecular assemblies; certain forms of cooperation were also essential. The evolutionary interplay of molecular competition and molecular cooperation reflected the need to process and utilize the first genetic information so as to stabilize and then improve it. It is impossible to re-create the actual stages of genetic improvement because enormous numbers of chance mutations were tested and discarded during early evolution. Nevertheless, one can now understand the natural laws that governed those stages. The laws can be tested in diverse ways: through experiments with bacterial viruses, through chemical studies of the components of nucleic acids and proteins and through comparative analysis of nucleic acids and proteins that have survived three or four billion years of molecular evolution.

THE EARTH BEFORE LIFE BEGAN

Before the drama of life could unfold the stage had to be set and certain minor actors had to take their places. The stage was somewhere on the primitive earth, which had a temperature not much different from what it is today. The composition of the earth's surface was also much as it is today if one considers the mere abundance of elements, but it was vastly different in the ways the elements were combined. Experiments have shown that almost any source of energy, such as lightning, shock waves, ultraviolet radiation or hot volcanic ash, would have led to significant conversions of early surface materials into a great variety of substances that would now be considered organic. The early solar system also included a very large amount of cometary and meteoritic material, which may have contributed substantially to the earth's surface. The effects of solar radiation on this ultracold material left over from the condensation of the solar system could have produced organic molecules as large as some biological polymers.

All conceptions of the 'primordial soup' from which life arose agree in that it included not only the particular sugars, amino acids and other substances that are now essential biochemical reactants but also many other molecules that are now only laboratory curiosities. It was therefore necessary for the first organizing principle to be highly selective from the start. It had to tolerate an enormous overburden of small molecules that were biologically 'wrong' but chemically possible. From this background the organizing principle had to extract those molecules that would eventually become the routinely synthesized standard monomers of all the biological polymers, and it had to link them dependably in particular configurations.

The total amount of potential organic material was immense. If the carbon now found in coal, carbonate rocks and living matter were uniformly distributed in all of the present ocean water, it would make a carbon solution as concentrated as a strong bouillon. Geophysical processes such as weathering, evaporation and sedimentation must have acted then as they do now to create a diversity of environments. Evidently at least one of these environments was suitable in temperature and composition for the origin of life.

The primitive soup did face an energy crisis: early life forms needed somehow to extract chemical energy from the molecules in the soup. For the story we have to tell here it is not important how they did so; some system of energy storage and delivery based on phosphates can be assumed. Nonmetabolic replenishment of the phosphate energy reservoir (perhaps by some kind of conversion of solar energy to chemical energy) had to last until a mechanism evolved for fermenting some otherwise unneeded components of the soup. Fermentation would have been adequate until the advent of photosynthesis provided a continuing energy source.

THE FIRST GENES

In cells genetic information is stored on DNA, transcribed onto messenger RNA and then translated into protein; in viruses the information is carried on either DNA or RNA strands. Both nucleic acids are threadlike molecules made up of nucleotides. Each nucleotide has three components: a chemical group called a base, a sugar (deoxyribose in DNA, ribose in RNA) and a phosphate group. The linked sugars and phosphates form the backbone of the molecule; the genetic information is encoded in particular sequences of bases. In DNA the four bases are the purines adenine (A) and guanine (G) and the pyrimidines thymine (T) and cytosine (C); in RNA uracil (U) takes the place of thymine. The bases are complementary, and they pair in accordance with specific rules: A pairs with T (or U) and G pairs with C. The complementarity is the basis of replication and transcription. In replication a strand of DNA or RNA serves as a template along which complementary nucleotides are assembled according to the base-pairing rules (by the various enzymes known as replicases and polymerases) to form a complementary strand carrying a duplicate copy of the information. In transcription a DNA sequence gives rise by a similar assembly process to a complementary strand of messenger RNA.

Knowing the chemical properties of DNA and RNA, what can one deduce about the identity of the first prebiotic information carriers? The deoxyribose nucleosides from which DNA is assembled are more difficult to deal with chemically than their ribose counterparts in RNA. Indeed, the synthesis of the monomers of DNA in cells proceeds via ribose intermediates, and DNA replication itself is initiated with RNA primers. In present organisms genetic information is processed by complex protein-RNA machinery. For such machinery to have evolved, the information carriers themselves must have had structural features that made them targets of recognition. Single-strand RNA can fold to form a great variety of three-dimensional structures, in contrast to the uniform double helix of DNA. In the present cellular machinery, wherever both functional and instructional properties are required, RNA is found. There is no reason to think it was otherwise during life's early stages. Nor is there any reason to think there was a process whereby information stored in any other form could have been transferred onto nucleic acids.

The search for the likely chemical identity of the first genes thus leads quickly to the base sequences of RNA. One can safely assume that primordial routes of synthesis and differentiation provided minute concentrations of short sequences of nucleotides that would be recognized as 'correct' by the standards of today's biochemistry: the sequences had the same bases, the same covalent bonds and the same stereochemistry, or spatial arrangement of chemical groups. These sequences were present, however, with myriads of others that would be regarded today as 'mistakes,' with different stereochemistry, misplaced covalent bonds and nonstandard bases. What was so special about the sequences that resembled today's RNA?

There is a simple answer. Those RNA strands with a homogeneous stereochemistry and with the correct covalent bonding in the backbone of the strand could reproducibly lead to stable secondary structures, or foldings of the molecule, as a result of the formation of hydrogen bonds between pairs of complementary nucleotides. This was an important advantage, making the strands more resistant to hydrolysis, the cleavage by a water molecule that is the ultimate fate of polymers in water solution.

The primitive RNA strands that happened to have the right backbone and the right nucleotides had a second and crucial advantage. They alone were capable of stable self-replication. They were simultaneously both the source of instruction (through the base-pairing rules) and the target molecules to be synthesized according to that instruction. Here at the molecular level are the roots of the old puzzle about the chicken or the egg. Which came first, function or information? As we shall show, neither one could precede the other; they had to evolve together.

The chemical species and processes of prebiotic times surely had a variety of features in common with present-day biochemistry. Sidney Fox and his colleagues at the University of Miami have shown, for example, that enzymatic functions can be exercised by 'proteinoid' polymers made essentially by warming a mixture of amino acids (the constituents of proteins). In addition to such primitive catalysts there were undoubtedly molecules that were receptive to stimulation by sunlight; there were lipids (fats) or lipidlike molecules that could form membranous structures and there were perhaps even polysaccharides, or sugar polymers, that were potential sources of energy. In short, a wealth of functional molecules had been created by nonliving, or 'nonorganic,' chemical paths.

Such functional molecules may have been important in the chemistry of a prebiotic soup. They could not evolve, however. Their accidental efficiency rested on nonaccidental structural constraints, such as favourable interactions with neighboring molecules or particular spatial foldings. If their efficiency was to improve, and if more functional variants were to be favoured over less functional ones, they would have to escape such structural constraints. Only self-replicative, information-conserving molecules could do so. We shall now discuss how the information content of such molecules can improve, how their complexity increases and how they drive out less functional variants.

SELF-REPLICATION

The virus Q_β, which infects the bacterium *Escherichia coli*, serves as a model system for studies of molecular self-replication. It has as its genome, or total hereditary material, a single-strand RNA molecule about 4,500 nucleotides long. Only part of the molecule constitutes the genetic message; the rest has a variety of functional (rather than informational) roles, such as serving as a target for specific recognition by enzymes. Some years ago Sol Spiegelman, who was then at the University of Illinois, isolated the Q_β replicating enzyme and showed that it is able to reproduce the virus RNA in a cell-free laboratory system to yield infectious copies. He also isolated from infected *E. coli* cells a noninfectious 'satellite' RNA molecule 220 nucleotides long that is replicated by the Q_β replicase with exceedingly high efficiency. The satellite RNA and other similar 'minivariants' serve in combination with Q_β replicase as convenient model systems for studies of RNA replication.

A typical experiment begins with a solution that includes magnesium ions, a low concentration of highly purified Q_β replicase and an energized form of the four RNA substrate monomers. The energized form consists of the nucleoside triphosphates *ATP*, *GTP*, *UTP* and *CTP*, in which the base and sugar are linked to a tail of three phosphate units. One of the four nucleosides (usually *GTP*) is labeled with a radioactive isotope so that the synthesis of new RNA can be followed. To initiate replication a certain amount of template RNA is added and the mixture is incubated.

In replication experiments done in 1974 by Manfred Sumper in our laboratory at the Max Planck Institute for Biophysical Chemistry in Göttingen something quite unexpected happened. When Sumper started incubation with more template than enzyme, he found a linear increase in RNA concentration, leveling off finally at a high concentration. This told us that all the enzyme molecules could simultaneously be occupied with templates being replicated; in spite of the steadily growing concentration of template, the concentration of active enzyme-template complexes was constant. Therefore new RNA continued to be synthesized at a constant rate.

A natural thing to do next was to reduce the amount of RNA template in the initial mixture. This resulted in shifting the linear growth curve parallel to itself to a later time. Successive reductions delayed the growth by an amount proportional to the logarithm of the template concentration. In other words, dilution from 10^6 to 10^4 template molecules per test tube shifted the growth curve by the same amount as dilution from 10^4 molecules to 10^2. This logarithmic relation clearly indicated that when enzyme is in excess of template, each newly formed RNA molecule can immediately find a free enzyme molecule. The template concentration is therefore amplified exponentially rather than linearly. This process works even for minute initial amounts of template; indeed it works with just one initial template molecule per test tube. (A method for cloning single molecules is based on this finding.)

Imagine our surprise when Sumper reported one day that even if not a single RNA template molecule was added initially, RNA was still produced, albeit only after much longer and more variable incubation times. The possibility that one of the enzyme molecules might have carried some RNA impurity into the mixture was eliminated by several procedures. The substrate nucleosides were subjected to conditions under which any polymer would have been totally degraded. The enzymes were purified and analyzed with all possible care. Single impurity molecules of the template were added deliberately in order to demonstrate that an entirely different mode of growth was caused by impurities. Finally we were convinced we had before us RNA molecules that had been synthesized de novo by the Q_β replicase enzyme. What was most puzzling, the de novo product had a uniform composition, which in each trial turned out to be similar to or even identical with Spiegelman's minivariant.

Comparative studies of chemical reaction rates soon showed that the mechanisms of template-induced synthesis and of template-free synthesis are quite different. The complicated template-induced mechanism was resolved into elementary steps in terms of which the results of our rate studies could be quantitatively matched by algebraic expressions. In the template-induced reaction one enzyme molecule is associated with one template molecule, and one substrate monomer is added at each step in the elongation of the replica; no cooperative interaction among substrate monomers can be observed. Template-free synthesis, on the other hand, does require the cooperation of at least three or four substrate monomers in the rate-limiting step of the reaction. Moreover, at least two enzyme molecules, each one loaded with substrate, participate in this step. One of the enzyme molecules apparently substitutes for the missing template by exposing bound substrate monomers to the polymerizing enzyme.

Spiegelman and Donald R. Mills of the Columbia University College of Physicians and Surgeons had determined the full sequence of the 220-nucleotide minivariant. On analyzing the sequence we saw that it could be represented as being composed (apart from 56 mutations and two insertions) of multiple copies of four tetramers and two trimers. The tetramers are CCCC and UUCG and their complements, GGGG and

CGAA; the trimers, *CCC* and its complement *GGG*, represent truncated versions of tetramers. The sequence *CCC,* had been identified by Sumper and Bernd-Olaf Küppers as the recognition site that must be present in all RNA's that interact specifically with Q_β replicase; *UUCG* is a base sequence that, in a different context (the translation of messenger RNA into protein), binds to one of the proteins that act as subunits of the Q_β replicase complex.

Does the discovery of the de novo synthesis of RNA violate the central dogma of molecular biology, according to which information can flow only from nucleic acids to proteins, and not the other way? The selection of these particuar tetramers and trimers clearly represented 'instruction' on the part of the proteins of the Q_β replicase complex. A wealth of possible sequences might have been composed of the tetramers and trimers, however, not just one, and a population of templates in the experiments comprises as many as 10^{12} molecules, only one of which needed finally to be amplified. Was natural selection at work rather than instruction by proteins?

THE ROLE OF SELECTION

Recently Christof Biebricher and Rüdiger Luce in our laboratory did a decisive experiment that answered the question. Their approach was determined by the special kinetics of de novo synthesis. They began by incubating a template-free mixture for a time that was long enough to amplify any templatelike impurity that may have been present but was too short for the formation of de novo RNA. The mixture was then divided into several isolated compartments where optimal conditions for de novo synthesis were maintained. The result was clear: Although each compartment had a uniform population of de novo products, the products differed from compartment to compartment. Further analysis revealed, however, that the different sequences were not completely unrelated.

Large fluctuations in the appearance times of different populations could be observed when compartments were incubated separately. The fluctuations reflect the probabilistic nature of a process in which the synthesis of a single molecule is the first, rate-limiting step. In contrast, template amplification always proceeds deterministically, with well-defined time constants, even if the reaction starts with only one template molecule or a few of them. Fluctuations in the rate of amplification average out during successive replications.

The early products appearing in the different compartments were clearly not yet optimized by any evolutionary process. Some were as short as 60 nucleotides, and still shorter ones had probably prevailed in the early stages of amplification. (An analyzable quantity of RNA consists of at least 10^{12} molecules. This number is approximately equal to 2^{40}, implying that 40 generations of amplification had passed before the products could be evaluated. In those 40 generations the most inefficient templates may well have been improved.)

Serial-transfer experiments, whereby growth can be prolonged through very many stages of amplification, quickly showed what the optimal products were like. They generally had lengths between 150 and 250 nucleotides. There was a definite, uniform final product for any set of experimental conditions, but there were as many different optimal products as there were different experimental conditions. One of the optimal products appeared to be Spiegelman's minivariant (which under the conditions of Sumper's experiments had shown up reproducibly). Other products of optimization were adapted to conditions that would destroy most RNA's, such as high concentrations of ribonuclease, an enzyme that cleaves RNA into pieces. Apparently the variant that is resistant to this degradation folds in a way that protects the sites at which cleavage would take place. Some variants were so well adapted to odd environments that they had a replication efficiency as much as 1,000 times that of variants adapted to a normal environment.

These observations leave no doubt that Sumper's results demonstrated de novo synthesis. The uniformity of the de novo products is seen to be a consequence of natural selection and not of faithful sequential instruction by the enzyme. And so the central dogma is saved, at least in its essence.

What is more important here is what the experiments reveal about Darwinian processes. Natural selection and evolution, which are consequences of self-reproduction, operate in the case of molecules as they do in the case of cells or species. What is truly surprising, and an important discovery indeed, is the high efficiency of the process of adaptation in such a simple self-reproduction system.

It may be objected that an enzyme such as Q_β replicase, a complex biological molecule, should not be present in an experiment designed to represent the prebiotic situation closely, even though the replicase was not a target of reproduction or evolution, but simply a factor in the environment. Fair enough! And that raises another important question.

TEMPLATE WITHOUT ENZYME

If the functioning of RNA as a template always required the participation of something as complex as Q_β replicase, prebiotic evolution would require optimization procedures beyond those based on the self-reproduction of RNA. It is therefore important to establish what kinds of self-reproduction and selection can take place in environments simpler than those that include well-adapted replicases. Then we can consider how instructed synthesis of proteins can originate on a Darwinian basis.

This question must be answered by experiment. Important clues can be cited from the recent work of Leslie E. Orgel and his colleagues at the Salk Institute for Biological Studies. Short polymers of the adenine nucleotide (oligo-A) form spontaneously when A monomers are exposed to templates that are long polymers of the complementary nucleotide U (poly-U), even when no enzyme or other catalyst is present. The length of the oligo-A chains averages five nucleotides and can be as much as 10. If lead ions are present as a catalyst, the yield is dramatically increased; in addition the successive monomers are for the most part (75 percent) linked to each other as they are in RNA: by a phosphate group that forms a bridge from the 3′ carbon atom of one sugar to the 5′ carbon of the next. If poly-C is the template in a 50-50 mixture of activated A and G monomers with lead ions present, the products have an overall 10:1 ratio of G to A; in other words, more than 90 percent of the base pairing is correct. When zinc ions are present, poly-C template and activated G monomers yield oligo-G chains of up to about 40 bases, and the fidelity is 20 times better than it is with the lead-ion catalyst. Has nature perhaps 'remembered' how replication started? Today's RNA polymerases all include zinc ions.

Orgel's data show that polymers rich in G and C offered special advantages in early evolution. They alone had sufficiently high copying fidelity in the absence of well-adapted replicases; they alone provide enough 'stickiness' in their base pairing to enable sizable messenger RNA's to be translated into functional proteins in the absence of ribosomes, the sites of translation in present-day cells. Kinetic and thermodynamic studies by Dietmar Pörschke in our laboratory have put these conclusions on a quantitative basis. G-C pairing proves to be about 10 times as strong as A-U pairing, so that complementary strands remain in contact with each other much longer when they are rich in G and C. Furthermore, the binding is strengthened cooperatively by neighboring pairs. From these data we have derived pairing rules for an evolutionary model that makes it possible to identify well-known RNA structures (such as the characteristic cloverleaf of transfer RNA's) as the evolutionary outcome of trial-and-error processes.

The essential conclusion of these enzyme-free studies is that RNA self-replication does indeed take place without the assistance of sophisticated enzymes. One can proceed to consider the evolutionary consequences of RNA self-replication without worrying about whether it really went on in prebiotic times. It did go on.

RNA QUASISPECIES

Suppose there were an endless supply of activated RNA monomers and the lifetimes of RNA's were infinite. What kind of self-replication would take place? Any RNA formed by noninstructed chemistry would be reproduced by template-instructed chemistry at a rate proportional to the current RNA concentration. The result would be exponential growth. Furthermore, even if only a single template were formed initially by noninstructed synthesis, there would soon be a host of different sequences because errors (point mutations, insertions and deletions) would inevitably be made in the course of replication. Hence in each generation there would be not only a larger number of RNA strands but also a greater variety of RNA sequences. What would happen then? Some of the mutants would be copied more rapidly than others or would be less susceptible to errors in copying, and their concentration would increase more rapidly. Sooner or later these faster-growing mutants would take over.

The same general picture emerges if monomers are supplied only slowly, so that the growing polymers have to compete for them, and if account is taken of the fact that a strand of RNA has a finite lifetime. Self-replication is a competitive process; the best competitor is the mutant sequence with the most favourable combination of copying fidelity, stability and replication rate. This is the basic line of thought along which one has to proceed in order to understand the self-replication experiments we described above, and on which our theory of self-replication is based.

The theory of competition in molecular self-replication is summarized in Appendix I (p. 28). The result of the competition is the 'survival' of the RNA sequence that is best adapted to the prevailing conditions, which we call the master sequence, together with a 'comet tail' of similar sequences derived from the master sequence by mutation. Although the values of the rate coefficients that pertain to primordial chemistry are not known, quantitative conclusions can nonetheless be drawn. One conclusion is that there is a threshold condition for the stable self-replication of a genetic message. Until circumstances allowed this threshold to be crossed no genetic message of any kind could survive.

One can estimate the maximum gene lengths available to prebiotic systems by inserting plausible values in the equation that determines the length of a gene in our model of self-replication. The maximum gene lengths range from 50 to 100 nucleotides, which is similar to the size of today's transfer RNA's. This is a reassuring result in that it is long enough for internal folding and stability, but at first it appears to be much too short a genetic message to encode a functional protein.

Before showing how these theoretical results are confirmed in the laboratory, let us consider just what it is that is selected in the RNA self-replication competition. In one sense it is the fittest of all the genes present (that is, the master sequence) because this sequence is present in the highest concentration. The master sequence is likely to constitute only a small fraction of the total gene composition, however. The number of mutants would be extremely large under prebiotic conditions because the chemical kinetics of most mutant sequences could not have been much different from the kinetics of the master sequence itself. Hence the result of the self-replication competition had to be the master sequence together with a huge swarm of mutants derived from it and from which it had no way of escape.

We call this entire mutant distribution a quasispecies. It is the quasispecies mutant distribution that survives the competition among self-replicating RNA's, and not just one master sequence or several equivalent ones that are the fittest genes in the distribution. The essence of selection, then, is the stability of the quasispecies. Violating the error-threshold relation is equivalent to destabilizing the quasispecies. The master sequence is then unable to withstand the accumulation of errors; the distribution starts to drift, and finally all information is lost.

The theoretical equations describing self-replicative competition have been tested and confirmed with data from cloning experiments on Q_β self-replication done by Charles Weissmann and his co-workers at the

University of Zurich. They measured short-time and long-time replication rates and studied the competition among mutant clones and wild-type Q_β RNA. Quantitative analysis of the data yielded values for copying fidelity and for competitive advantage in agreement with theory. The experiments showed that even with a highly evolved replication apparatus, organisms contend with less-than-perfect fidelity of replication by limiting gene lengths and by surviving with quasi-species distributions rather than with unique genes.

ERROR, GENOTYPE AND PHENOTYPE

We mentioned above the energy crisis that had to be overcome in the first stages of biogenesis. We now discuss an obstacle that played an even larger role in the evolution of life: an information crisis.

The first Darwinian molecular systems owed their self-replicating ability to inherent physical forces that brought about the formation of complementary base pairs. The error threshold set a length limit of about 100 nucleotides, attainable only by RNA sequences rich in G and C nucleotides. This limit was eventually overcome by the development of a capability for the translation of genes into protein, and thus of enzyme machinery that reduced the error rate enough to make possible gene lengths of up to several thousand nucleotides. This new barrier is still reflected in the limited gene lengths of present single-strand RNA viruses, even though the viruses are a much later evolutionary development.

Further extension of gene length was possible only with the appearance of mechanisms for detecting and correcting errors. The distinction of right from wrong could then be made if the newly formed daughter strand remained associated with its parental template, in which case 'wrong' could be identified chemically as 'unpaired.'

All of this became possible when double-strand DNA appeared on the scene. DNA polymerases are equipped with proofreading and error-suppressing functions so effective that they allow gene lengths to be extended to millions of nucleotides. Lawrence A. Loeb of the University of Washington School of Medicine has shown that if a DNA polymerase is unable to carry out error-correcting functions, it has the same replication fidelity as an RNA replicase, namely a value between .999 and .9999.

The invention of DNA made possible the formation of individualized cells in which the division of the cell is synchronized with the replication of its genetic material. Now, however, a new information crisis had to be overcome: high-fidelity replication narrowed the opportunities to provide variability by means of point mutations. The new barrier was overcome by the development of recombinative processes, including eventually sexual reproduction, which grafted Mendelian genetics onto self-reproduction, the basis of Darwinian systems.

The first information crisis could be overcome only by the organization of a self-optimizing enzymatic replication machinery based on a stable quasispecies. This evolutionary jump required translation of RNA information into a new language: a functional one, namely proteins.

To encode even the most primitive translation apparatus certainly required far more than the 100 or so nucleotides that could be reproducibly stored in one master sequence. However the first protein machinery arose, it required more information than a primitive molecular Darwinian system was able to provide. More information could be stabilized only with the cooperation of differentiated genes, mediated and regulated by their very translation products.

If in the expanded information system the products of translation are also targets of evolution, a new problem arises. Selection has to act on the information content of the nucleotide sequence: on its genotype. Evaluation for selection, however, has to take place at the level of the gene product's function: the phenotype. This genotype–phenotype dichotomy demands that the system feed back information to its own genes, a procedure called second-order autocatalysis. It is second-order because in order to reproduce itself the information carrier needs both the information supplied by the template and the machinery encoded in the template. We have named double-feedback loops of this kind hypercycles. The term includes a large class of higher-order autocatalytic mechanisms. They exhibit a particular kind of temporal behavior different from that of other Darwinian systems.

HYPERCYCLES: QUASISPECIES COOPERATE

Hypercyclic coupling operates today when an RNA virus attacks a cell. If viral RNA were just another template in the replication environment of the host cell, it would not be able to outgrow other host templates. What it does instead is specify information for a replication machine that is highly selective for the viral RNA itself. Most parts of the machinery are provided by the host, but the specific hypercyclic linkage ensures the success of the virus's attack.

A simple example will explain the basic operating principle of a hypercycle. Suppose RNA sequence 1 codes for an enzyme 1 that helps to catalyze the self-replication of RNA sequence 2. Sequence 2 in turn codes for an enzyme 2 that helps to catalyze the self-replication of RNA sequence 1. What happens? Sequence 1 needs enzyme 2 for its self-replication and sequence 2 needs enzyme 1. Therefore neither sequence can afford to outcompete the other for the available supply of RNA monomers; the two sequences are forced to cooperate. Depending on the rates of the numerous catalytic steps, wide ranges of concentrations may prevail, but as long as there is interdependence only an unusual 'fluctuation catastrophe' or a major change in chemical conditions can extinguish an existing hypercycle. Templates and enzymes function together in self-replication, with the protein product of the RNA serving as a replicase, as an activator of a replicase or as some other control device enhancing the speed and accuracy of self-replication.

The kinetic behavior of such coupling into hypercycles has been investigated in detail and it has been shown that hypercycles are the only functional networks that can exceed the error threshold for stable quasi-species. Hypercyclic growth proves to be explosive when compared with first-order autocatalytic growth in a system having comparable rate co-efficients. The consequences of hypercyclic growth in selection are even more strikingly different.

Life could not have originated with such a simple hypercyclic scheme as the one described above. The first catalytic couplings must have been weak and complex, and the number of genetic participants (members of RNA quasispecies) and functional participants (primitive enzymes) very large. The hypercyclic principle itself was nonetheless simple: enforced cooperation among otherwise competing genes allowed their mutual survival and regulated their growth. It also made possible a more refined kind of evolution than that open to quasispecies alone.

In a quasispecies Darwinian competition evaluates the fitness of each mutant RNA according to its rate and accuracy of self-replication and its stability. When quasispecies are locked into a hypercycle, however, new criteria come into play. First, evaluation of the 'target function' of each quasispecies becomes crucial: those sequences are fittest that are best able to get themselves replicated quickly and accurately by the enzyme responsible for their replication. Second, the continual introduction of new mutant RNA sequences means that new catalytic couplings are constantly being tested. The structure of the hypercycle evolves whenever new catalytic couplings are discovered to be advantageous.

The hypercycle principle also shares an evolutionary disadvantage with the quasispecies. Both quasispecies competition and hypercyclic cooperation evaluate only phenotypic properties of the RNA's: their stabilities and reaction rates. If a way could be found to evaluate the genotypic properties of the RNA sequences—the genetic messages themselves—the quality of the enzymes resulting from translation of the sequences would be subject to improvement through natural selection. We see only one possible way: by putting the hypercyclically organized quasispecies distributions into compartments that could then evolve through Darwinian competition.

The transition from a single quasispecies distribution to the hypercyclic organization of many distributions probably came about smoothly rather than at a stroke. Primitive translation mechanisms in a quasispecies distribution sooner or later gave rise to proteins that were more helpful to self-replication than the miscellaneous proteins in the soup. At first the help must have been more or less uniform throughout the quasispecies distribution, but as preferences among translation products became more pronounced, cross-assistance between sequences became immensely more probable than self-feedback. The initially complex mesh of interactions became more and more distinct as the advantages derived from more specific catalysis had effect. Finally the differences among the various template-catalyst interactions became so great that each enzyme had a particular catalytic role. At this point the original single quasispecies distribution had diverged to form a set of distinct quasispecies distributions, and the first hypercycle was in operation. Hypercycles arose as naturally and continuously as quasispecies did; they arose under the force of natural law.

COMPARTMENTATION

Life now is everywhere cellular. Why? Among the obvious advantages of cellular organization are protection from fluctuations in the external environment and maintenance of concentration gradients, but such advantages do not explain the origin of the cellular organization principle. Cellular organization was needed because it was the only way to solve the one problem of information processing in evolution that self-replicative competition and hypercyclic cooperation were unable to address: the evaluation of the information in genetic messages.

Organization into cells was surely postponed as long as possible. Anything that interposed spatial limits in a homogeneous system would have introduced difficult problems for prebiotic chemistry. Constructing boundaries, transporting things across them and modifying them when necessary are tasks accomplished today by the most refined cellular processes. Achieving analogous results in a prebiotic soup must have required fundamental innovations.

Darwinian competition in a quasispecies distribution was based on selection according to the chemical kinetics of the sequences; what the sequences 'meant' played no role. The meaning of the message could not be ignored when hypercyclic organization of enzymes and RNA's came into play, since the meaning governed the strength of the coupling. The one-directional character of the cyclic coupling, however, still excluded any feedback that would allow genetic meaning to be evaluated and so make possible selection of the best information. In a hypercycle as in a quasispecies only the target function of an RNA (its affinity for the proteins that replicate it) is evaluated, not the genetic information the RNA encodes. A hypercycle in solution cannot select for its translation products, whether those products are advantageous or disadvantageous.

We see only one answer to the problem of evaluating the quality of the information in early genes. The answer was to provide a means of breaking up the homogeneity of the soup; it was to compartmentalize the evolutionary process. Once the mutational events in one compartment were made independent of contemporary events in other compartments, a way had been found to improve genetic information; the way, of course, was Darwinian evolution. Compartments that were fitter than others could be selected on the basis of their total performance, including the possession of better genetic information. As long as there was a way to pass on genetic information from one generation of compartments to the next, the evolution of the total information content was assured.

Now the flow of logic governing prebiotic evolution is complete. To attain any information stability at all required the self-replication of short RNA sequences. Darwinian competition among the mutant sequences led to a single quasispecies distribution as the potential product of evolution. Then hypercyclic organization arose among the mutant sequences and allowed many quasispecies distributions to coexist in the same soup. This expanded the amount and the variety of information present far beyond what had been allowed (by limited copying fidelity) in a single primitive gene, but it did not provide any opportunity for the information to be evaluated on the basis of its function in a competition that would lead to evolutionary improvement. That opportunity was provided by compartmentation and subsequent intercompartmental competition.

Compartmentation by itself could not have provided all that is logically necessary for the origin of life. In a compartment the problems of limited copying fidelity and of competition among self-replicating genes still had to be confronted. The only way so far discovered for maintaining enough genetic information to code for a minimal starting amount of enzymatic function is hypercyclic organization. Hypercycles and compartments addressed two independent problems of prebiotic evolution. Hypercycles provided for the stable coexistence of a variety of self-replicating genes and thereby solved the first information crisis. Compartments provided a way to evaluate and thus improve the genes' information content. In other words, compartments dealt with the genotype-phenotype dichotomy.

LIFE IN THE TEST TUBE?

If one can really deduce the natural laws that operated to create life on the earth, why not just assemble the necessary materials and re-create life in a test tube? Anyone attempting such an experiment would be seriously underestimating the complexity of prebiotic molecular evolution. Investigators know only how to play simple melodies on one or two instruments out of the huge orchestra that plays the symphony of evolution. The investigator substitutes for a single instrument much as an amateur musician does with a 'music minus one' record.

In this article, for example, we have described some test-tube experiments on molecular evolution that showed how RNA templates with quite sophisticated phenotypic properties originate when the necessary enzymatic machinery is present as an environmental factor. The next step is to find out how such machinery evolved in the first place. We would like to make all its components and test them experimentally.

What amino acids were components of the first proteins synthesized by RNA instruction? In the present-day genetic code each of the 64 (4^3) possible triplets of the four RNA nucleotides is a 'codon' that tells the translation machinery to add a particular one of the 20 amino acids to the protein chain (or to start or stop the translation process). That is surely much too complex a system to have been initiated in one step by natural causes. Was there once a more primitive precursor code? What was its structure? These questions have to be addressed before experiments on the self-organization of translation can be designed.

Primitive proteins must have been made from fewer amino acids, and so an initial one- or two-character code would have been sufficient to assign them all. There is no chemically simple way, however, to change from a singlet or doublet codon to a triplet one because in such a transition all existing messages would be nonsense until they were completely recoded. The genetic code must therefore have been based on a triplet frame from the start. What was that frame?

Prebiotic translation imposed on the code some tasks that today's translation mechanism solves by refined measures not connected with the code itself. At first the code itself had to establish the direction of readoff and the punctuation of the message by defining a 'reading frame.' In 1976 F. H. C. Crick, Sydney Brenner, Aaron Klug and George Pieczenik of the Medical Research Council Laboratory of Molecular Biology in Cambridge proposed that directionality and framing were initially fixed by translating only triplets that had the sequence *RRY*; *R* signifies a purine nucleotide (*G* or *A*) and *Y* a pyrimidine (*C* or *U*). They also noted that *RNY* triplets would serve the same purpose, with *N* signifying any nucleotide. Directionality and punctuation are established by *RNY* just as well as by *RRY*, and with *RNY* the same frame pattern would apply to both the template strand and its complementary daughter strand.

EVOLUTION OF THE CODE

Are there any clues in the present code and machinery as to whether it arose from frames such as *RRY* or *RNY*? The creation of a computer archive of sequence comparisons by Margaret Oakley Dayhoff and her co-workers at the National Biomedical Research Foundation has made it possible to undertake large-scale searches for genetic relations among biological polymers, and in particular to construct phylogenetic trees showing relations among homologous proteins or nucleic acids in different species. Transfer RNA's are particularly appropriate for such analysis with respect to the question of origins. Their function is to adapt each amino acid to its codon. Given such a key role, their structure might still reflect how the assignments of codons to amino acids came about.

In searching for primitive sequences by analyzing present ones it was not enough simply to process a lot of sequence data with a computer program designed to find the optimal phylogenetic tree. Analytical criteria to determine the degree of 'treelikeness' in a set of sequence data had first to be formulated and confirmed, and this was done through topological analyses in cooperation with Andreas Dress of the University of Bielefeld.

When these criteria and programs like Dayhoff's were applied to the analysis of all currently known transfer-RNA sequences (about 200), two fascinating conclusions emerged. First, the sequences of a given transfer RNA (for example the one that mediates initiation of translation) in all species studied do appear to be related to one another in a treelike pattern that reflects only small evolutionary divergence compared with the changes in other biological polymers. Apparently this particular early information has survived later stages of evolution quite well. Second, the sequences of different transfer RNA's in a single organism reflect divergence from a common ancestor, but they seem not to be related to one another in a treelike manner, at least not in the two organisms (*E. coli* and yeast) for which we have enough sequence data to make the analysis statistically meaningful.

The sequences seem instead to represent a mutant distribution similar to that of a quasispecies. In a kind of back-extrapolation through the quasispecies' history to the time of the origin of translation, the analysis identified what appeared to be ancestors of modern transfer RNA's and led to two important inferences about them. They were much richer in G and C than in A and U, and in their master sequences (which were determined by assigning the most prevalent base at each position) they showed a clear reverberation of a triplet pattern of the form RNY.

One can look for ancient genetic information in other places: wherever there is evidence that selection pressure and genetic drift have not yet reduced the 'remembrance' of ancestral sequences to below the noise level. John Shepherd of the University of Basel recently applied a new method of computer-aided sequence analysis suitable for long genetic messages. His method measures the distance between repetitions of characters or groups of characters along a sequence. Ancestral information can be distinguished from later modifications. His first conclusions, drawn from studies of several DNA viruses and genes of bacteria and higher organisms, are again that there is in these modern genes still a memory of ancient sequences and that the triplet RNY dominated those sequences.

The stability of the G-C pairing strongly suggests that the initial RNY code must have been limited to the four GNC codons. The present assignments of those codons are GGC = glycine, GCC = alanine, GAC = aspartic acid and GUC = valine. Simulations of primordial chemistry done by Stanley L. Miller of the University of California at San Diego suggest that these amino acids were among the most prevalent ones in the primordial soup. If that is a coincidence, it is certainly a suggestive one.

We have crossed the threshold of being able to have confidence in our ability to reconstruct ancestral RNA and protein sequences. On the basis of this information we are beginning to reconstruct and resynthesize ancestral sequences, both of proteins and of RNA's, and to test their interplay in a continuous-flow reactor, or what amounts to an evolution machine.

If the first proteins were indeed made of the four amino acids we mentioned, they had a negative electric charge. In general such amino acids would not readily associate with a negatively charged RNA species unless specific forces stabilize a particular interaction. Claude Hélène of the University of Orléans has shown that there is a strong specific interaction between the carboxylate (COO-) groups of such amino acids as aspartic acid and the G nucleotides of RNA. Hence particular sequences can indeed build up patterns for specific contacts, which then may be stabilized with the help of metal ions. The first specific catalysts for replication and translation were probably just such structures, mediating specific contacts and thereby supporting weak chemical functions.

All such functions must have been recruited from the information of an initial quasispecies, the mutants of which finally differentiated when they organized themselves into hypercyclic functional linkages. The principles guiding the evolution of such an organization have been formulated and experimentally verified. Now what remains to be discovered is just what the favourable molecular structures were.

Manfred Eigen, William Gardiner, Peter Schuster and Ruthild Winkler-Oswatitsch prepared this article at the Max Planck Institute for Biophysical Chemistry in Göttingen, where Manfred Eigen is director of the Department of Biochemical Kinetics. William Gardiner is Professor of Chemistry at the University of Texas at Austin and Peter Schuster is director of the University of Vienna's Institute for Theoretical Chemistry and Radiochemistry.

The origin of genetic information:
Appendix I
The quasispecies model

Prebiotic RNA chemistry provided an environment for Darwinian evolution: populations of self-replicating species (RNA strands with different sequences) competed for the available supply of 'food' (energy-rich monomers). The continuous generation of mutant sequences, some of them having advantageous properties, forced evolutionary reevaluation of the fittest species. A quantitative theory of this molecular Darwinian competition has been developed.

Let the number of nucleotides in any sequence i be N_i and let each nucleotide position by identified by a subscript p that can have any value from 1 to N_i. Let the probability that the nucleotide at position p in sequence i gets copied correctly during self-replication be q_{ip}; then $(1 - q_{ip})$ is the error rate for that position. The symbol q_{ip} therefore describes the quality, or copying fidelity, of replication at position p of sequence i. The probability Q_i that an entirely correct sequence i will result from a replication is the product of all the single-nucleotide copying fidelities:

$$Q_i = q_{i1} \times q_{i2} \times \ldots \times q_{iN_i} = \bar{q}_i^{N_i},$$

where $\bar{q}_i$ is the geometric mean of the copying fidelities in sequence i.

Sequence i can survive successive replications only if copying errors do not accumulate. This requires the sequence to be superior in net growth to the average of its competitiors by a competitive-advantage factor S_i. Furthermore, i can be selected only if a survival condition called the error threshold is satisfied; the threshold is Q_iS_i, and for sequence i to survive it must be greater than 1.

Net growth is governed by the equation that tells how x_i, the fraction of all the sequences present that are exact copies of sequence i, changes with time. Major causes of change in x_i are error-free replication of i and erroneous replication of closely related sequences, collectively designated j, which can give rise to i by mutation. Taking both contributions into account gives the rate of change of x_i:

$$(W_{ii} - \bar{E})\, x_i + \text{sum of } W_{ij}x_j.$$

In this expression W_{ii} is the rate of correct replication of sequence i and $\bar{E}$ is the average rate of excess production (excess of replicative gain over all losses) of all sequences present; both rates are expressed on a per-copy basis. W_{ij} is the rate of production of sequence i by erroneous copying of sequence j, given per copy of sequence j. In the equation the contributions from all the sequences designated j are summed. Hence the first term is the rate at which sequence i competes with other sequences and the second term is the rate at which i is produced by mutations of other sequences.

These rates determine how any arbitrary starting set of sequences proceeds to organize itself. The first term can be positive or negative depending on whether W_{ii} is greater than or less than the average excess production $\bar{E}$. If W_{ii} is greater, x_i grows; if it is less, x_i steadily decreases until sequence i dies out or is produced only by mutation. The decline of sequences for which W_{ii} is less than $\bar{E}$ has the effect, however, of increasing the average excess production rate $\bar{E}$. This makes it ever more difficult for surviving sequences to satisfy the requirement that W_{ii} be greater than $\bar{E}$, and hence to grow. The self-replicative competition resembles a high-jump competition in which the bar is raised until only one competitor survives. In the molecular competition, however, a single winner never emerges. Because of the mutation terms $W_{ij}x_j$ the strongest competitor constantly produces mutant sequences with which it must continue to compete. In the steady state that is eventually reached the best competitor, designated the master sequence m, coexists with all mutant sequences derived from it by erroneous copying. We designate this distribution of sequences a quasispecies.

This analysis shows that Darwin's principle of natural selection is not of an axiomatic nature but derives from the set of physical conditions that pertain to self-replication. The end result of selection, the quasispecies, is stable until mutation happens to produce a new sequence with a growth rate higher than that of the existing master sequence (or until a change in environmental conditions has an equivalent effect). When that happens, the new 'fittest' sequence proliferates until it (together with its mutants) takes over and the old quasispecies disappears.

Quantitative results describing RNA quasispecies have been derived. For example, the maximum length of a master sequence (the number of nucleotides) is

$$\frac{2.3 \log S_m}{1 - \bar{q}_m}.$$

Longer master sequences could not surmount the error threshold, that is, $Q_m S_m$ could not be greater than 1.

This discussion summarizes the main results of mathematical investigations carried out by our group and later by B. L. Jones, R. H. Enns and S. S. Ragnekar of Simon Fraser University in British Columbia and by C. J. Thompson and J. L. McBride of the University of Melbourne.

The origin of genetic information:
Appendix II
The hypercycle model

When RNA-instructed synthesis of proteins became a factor in evolution, there arose a new kind of dynamical interaction among molecules. The characteristic features of the interaction can be understood by means of a 'topological' analysis, in which only qualitative and not quantitative conclusions are reached.

Consider a set of several different master sequences with their mutant distributions, each master sequence and distribution being (in the absence of the other ones) a stable quasispecies. The total information content of all the master sequences exceeds the limit established by the error threshold for one master sequence alone. For such a set to be stable and retain this total amount of information three conditions must be met:

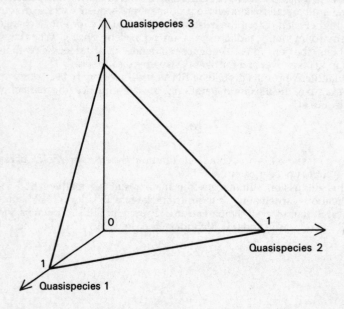

(1) Each quasispecies itself must remain stable, that is, each master sequence must compete successfully with its mutants, so that errors do

Quasispecies 3

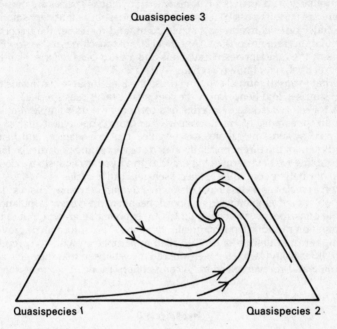

Quasispecies 1   Quasispecies 2

not accumulate; (2) the different master sequences, with different selective values, must tolerate one another because of mutual catalytic couplings; (3) the coupled set must remain stable, regulating the populations of its members and competing as a set with alternative sets.

The topological analysis begins with the definition of a composition space in which each coordinate axis represents the fractional population of a quasispecies (the number of RNA strands belonging to the quasispecies divided by the total number of RNA strands present). For three quasispecies a three-dimensional composition space is defined. A state of the system is characterized by three fractional-population values and is represented by a point in the three-dimensional composition space. Because the three fractional populations are all positive and their sum is 1, this point must lie somewhere on the equilateral triangle whose vertexes are at 1 on each axis of the composition-space coordinate system.

The equilateral triangle is called the unit simplex. If more than three quasispecies are considered, the unit simplex is a geometric figure of higher dimension. The triangle's corners represent states of the system comprising only one quasispecies, the edges of the triangle represent states with two quasispecies and the interior of the triangle represents states in which all three quasispecies are present.

A temporal sequence of states is described by a 'trajectory,' a curve on the unit simplex. Methods are available for discovering the qualitative nature of trajectories without solving the dynamical equations, which cannot be solved analytically if more than two quasispecies are present. There are various possible classes of trajectories. In a stationary solution the trajectories converge to a stable point, and thereafter the fractional populations remain constant. Another class of trajectories reflects oscillations of the quasispecies populations and a third class reflects a kind of irregular behavior known as chaos.

What forms of mutual coupling cause the trajectories to stay inside the unit simplex, and hence imply the coexistence of all quasispecies?

A trajectory leading to an edge or a corner of the unit simplex implies the disappearance of one quasispecies or more. Topological analysis of coupled systems in general has revealed that coexistence, and hence fulfillment of the three conditions stated above, requires a particular kind of coupling we have termed hypercyclic. In a hypercyclic system a closed loop of catalytic couplings connects self-replicative cycles.

Hypercycles as a class have distinctive dynamical characteristics. The growth rate of a hypercycle is proportional not to the current populations of the quasispecies, which would lead to exponential growth, but to the population raised to a power greater than one. This autocatalytic growth of higher than first order can be called hyperbolic growth. Hypercycles also differ from Darwinian self-replicative systems in that they give rise to 'once-and-forever selection.' A competition among hypercycles can be

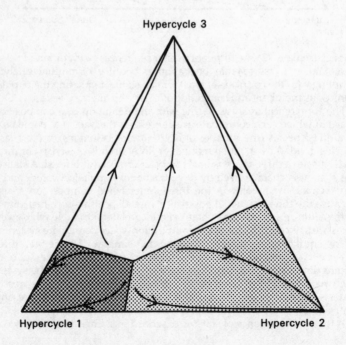

Hypercycle 3

Hypercycle 1

Hypercycle 2

analyzed topologically in the same way as a competition among quasi-species. The composition space gives the fractional populations of the competing hypercycles, and the trajectories lie on a hypercycle unit simplex. It turns out that all trajectories lead to the vertex of the area of the unit simplex in which the competition began. In other words, any competition among hypercycles leads to the survival of only one hypercycle.

Once-and-forever selection implies that a hypercycle, once established, cannot be displaced by a competitor appearing in small numbers even if the competitor is more efficient. This reflects the fact that the selective value of a hypercycle depends on its population. That is not the case for Darwinian systems, where a single advantageous mutant molecule can outgrow an established population. Hypercycles can evolve, however, by optimizing their internal linkages as a result of replacements, insertions and deletions in the information-carrying molecules. Because hypercycles provide complete mutual control of the coupled populations they grow coherently.

Hypercycles appeared in evolution with the origin of translation, which introduced a new requirement: it became necessary to feed back an evaluation of the quality of translation products to the genes that encoded those products. Presumably hypercycles still exist today as features of viral infection processes.

Hypercycles and the origin of life

JOHN MAYNARD SMITH

PERHAPS the most difficult step to explain in the origin of life is that from the replication of molecules (RNA for example) in the absence of specific proteins, to the appearance of polymerases and other proteins involved in the replication of RNA and themselves coded for by that RNA. Suppose we start with a population of replicating RNA molecules. Without specific enzymes the accuracy of replication is low and hence the length of RNA which could be precisely replicated small. Before replication can be reasonably accurate, there must as a minimum be a specific polymerase, as well as synthetases and tRNAs, which in turn implies an RNA genome of considerable length. Thus, even if one supposes an initially very limited set of codons, one cannot have accurate replication without a length of RNA of, say, 2,000 or more base pairs, and one cannot have that much RNA without accurate replication. This is the central problem discussed in a series of papers by Manfred Eigen and Peter Schuster[1] proposing the 'hypercycle' as a necessary intermediate stage.

First, imagine a population of replicating RNA molecules, lacking genetic recombination, but with a certain 'error' or 'mutation' rate per base replication. Very roughly, if more than one mutation occurred per molecular replication, the population would come to consist of a random collection of sequences. But if less than one mutation occurred per replication, and if one sequence was 'fitter' than others in the sense of being more stable and/or more easily replicated, then a population would arise whose sequences were centred around this optimal one. Such a population Eigen and Schuster call a molecular 'quasi-species'. If the mutation rate per replication was much less than one, and if fitness differences were large, most molecules would have the optimal sequence. As the mutation rate increased, the proportion of the population with the optimal sequence would fall, until a critical mutation rate was reached, above which sequences would be random.

The replication of RNA molecules in the absence of specific replicases has not yet been achieved in the laboratory; it may initially have depended on the presence of nonspecific random sequence polypeptides. At this stage, the error rate would depend solely on the energy levels of base pairing between G and C, and between A and U, and would be such that the maximum length of a quasi-species would be 10–100 base pairs; even this length would require the RNA to be G–C rich. In contrast, RNA replication by the enzyme coded for by the RNA phage Q_β is accurate enough to permit a genome of 1,000–10,000 base pairs, which would be sufficient to code for a primitive protein-synthesising machinery. The next stage in the evolution of accuracy of replication, achieved by prokaryotes, was the recognition and correction of mismatches in DNA replication, a process which depends on recognising which is the old strand and which the new. With such proof-correcting, a genome of at least 10^6 base pairs, and perhaps 10^8–10^9 base pairs, can be maintained. It is the gap between the 10–100 bases, without specific enzymes, and the 1,000–10,000 bases, with specific enzymes, that Eigen and Schuster aim to bridge.

To understand their proposal, imagine that you wish to replicate the message GOD SAVE THE QUEEN. Counting 5 bits per letter, this is a total of 75 bits, and a maximum of 25 bits per word. Suppose the error rate was, say, 1/50 per bit. You start with a population of messages, copy each of them with that error rate, and then discard half of the population, applying a rule of selection so that messages with many errors are most likely to be discarded. Despite this 'natural selection', the population would steadily accumulate errors. The mutation rate is too high.

Suppose, therefore, that after each replication you selected word by word, discarding half the quasi-species GOD, and similarly for each other word. Since no word contains as many as 50 bits, you could now maintain a meaningful message. However, there is an important reason why this is *not* an adequate model of the replication of a set of RNA molecules. Thus to maintain complete messages you would have to ensure in each generation that you selected equal numbers (or approximately so) of each word. But suppose the words competed, as RNA molecules would compete, for substrates. Then before long only one quasi-species would remain—GOD only, or THE only, if short words replicate faster.

Hence if the words are strung together into a single selected message (analogous to a single RNA molecule), the message is too long to replicate; if they are replicated separately, competition between the words destroys the message. You can escape from this dilemma by arranging the words in a 'hypercycle', as in Fig. 1. In this cycle, the *rate* (not the accuracy) of

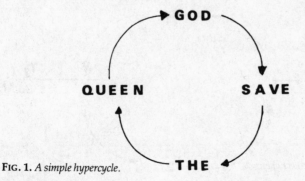

FIG. 1. *A simple hypercycle.*

replication of SAVE is increased by the number of GODs, of THE by the number of SAVEs, of QUEEN by the number of THEs, and, coming full circle, of GOD by the number of QUEENs. Each word, or quasi-species, is selected independently, according to its own fitness. This is a hypercycle. It can be shown that only by linking together a set of replicating quasi-species in this cyclical way can the stability of the whole be maintained.

A possible way in which a simple 'two word' hypercycle might be realised in molecular terms is shown in Fig. 2. (Eigen and Schuster would not insist on this particular realisation; it is intended only to show the kind of thing they have in mind). It is necessary to describe a cycle in some detail, so as to be able to discuss the difficulty which arises in explaining the further evolution of such cycles. I_1 and I_2 are two RNA quasi-species. Both the + and − strands must be good replicators. RNA molecules have a phenotype, because they fold up, and folding patterns affect both survival (resistance to hydrolysis) and replication rate. Since both + and − strands must replicate, they are likely to have similar folding patterns. Further, I_1 and I_2 may be descendants of the same ancestral quasi-species, in which case they will resemble one another.

In this particular realisation, the − strands T_1 and T_2 are supposed to be 'adapters' or precursors of tRNA, coupling with specific amino acids, and having an anticodon loop. The + strands M_1 and M_2 are 'messengers', composed of only two kinds of codon, and coding for two proteins E_1 and E_2 which act as replicases for I_2 and I_1 respectively. Thus the model assumes that some kind of translation is possible without synthetases or ribosomes, and that the code is arbitrary from the start (that is, that there is no chemical constraint on which triplet codes for which amino acid). It is not an essential feature of the argument that messenger and transfer molecules be the + and − strands of the same quasi-species, but if they were not, then a hypercycle of more than two elements would be needed.

One difficulty in conceiving how such a cycle might arise is that it requires that each of two 'messengers' (or of more than two in the case of a longer cycle) should programme a replicase for the other. Eigen and Schuster argue that this becomes less implausible if it is remembered that I_1 and I_2 may be descended from members of the same molecular quasi-

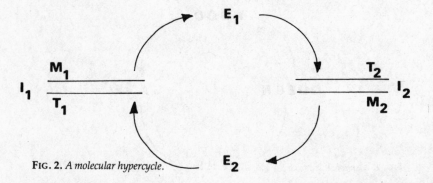

FIG. 2. *A molecular hypercycle.*

species. Initially, the members of a quasi-species would be sufficiently alike to share the same replicase. Hypercyclic organisation could then arise by the gradual differentiation of a single quasi-species into two or more.

I now digress to give the authors' views on the origin of the code. They argue that primitive RNA molecules must have been G–C rich (for reasons already mentioned). Further, in the absence of ribosomes the message must have been one which could only be read 'in frame'. These two conditions imply that the earliest messages consisted of strings either of GNC codons or CNG codons, where N stands for any base. The authors use an argument based on wobble to prefer the former codon type. Hence, from arguments based on the stability, replication and translation of nucleic acids, they conclude that the first two codons were GGC and GCC, followed by GAC and GUC. This conclusion agrees nicely with the likely abundances of amino acids in the primitive oceans. By far the most abundant amino acids in simulated prebiotic synthesis[3] are glycine and alanine (today coded by GGC and GCC), the next commonest being aspartic acid (GAC) and valine (GUC).

Returning to the main theme, the following question arises. Given that a hypercycle ensures the accurate replication of a larger total quantity of information, how will it evolve further? Consider the particular realisation in Fig. 2. There are three kinds of mutation which might occur in I_1 (and a similar set in I_2):

(1) Mutations making I_1 better (or worse) at replicating, for example by becoming a better target for E_2. (I_1 might also become a better target for E_1; if this process went too far the hypercycle would be disrupted).
(2) Mutations making E_1 a better (or worse) replicator of I_2.
(3) Mutations making T_1 a better adapter.

If there is not compartmentalisation, only mutations of type (1) would be incorporated by selection. Each quasi-species in the hypercycle would evolve independently. There would be no selection favouring mutations of types (2) and (3), although such mutations would be needed before the speed and accuracy of replication would improve sufficiently to permit the genetic information to be united in a single 'chromosome'. There is a natural analogy here to an ecosystem. Imagine an ecosystem consisting of grass, antelopes and lions. The more grass there is, the more rapid is the multiplication of antelopes. Similarly, antelopes encourage the multiplication of lions, and (by some stretch of the imagination) lions, by fertilising the ground, encourage the growth of grass. The point of the analogy is this. Natural selection will favour mutations in grass which increase the fitness of individual grass plants, but will not favour changes in grass making it more edible to antelopes.

How then can a hypercycle evolve characteristics which favour the growth of the cycle as a whole, rather than merely its constituent parts? So long as there is no compartmentalisation, it cannot. For natural selection to act, there must be individuals. In the present context, this seems to require that the compartments of a hypercycle be enclosed in a membrane to form a proto-cell. If these proto-cells grew and divided—by some kind of budding or fission—at a rate which increased with the growth rate of the enclosed hypercycle, then selection would favour mutations making the hypercycle more efficient. In the example of Fig. 2, mutations of types (2) and (3) would be favoured.

Clearly, these papers raise more problems than they solve. Their merit is that they put in sharp terms the problem raised by the relatively inaccurate replication of nucleic acids, and, in the hypercycle, they suggest a way in which a relatively complex structure could be replicated despite this inaccuracy.

1. Eigen, M. & Schuster, P. *Naturwissenschaften* **64**, 541 (1977); **65**, 7 (1978); **65**, 341 (1978).
2. Crick, F. *et al. Origins of Life* **7**, 389 (1976).
3. Miller, S. L. & Orgel, L. *The Origins of Life on Earth* (Prentice-Hall, 1973).

John Maynard Smith is in the School of Biological Sciences, University of Sussex

This article first appeared in *Nature* Vol. **280**, pp. 445–446; 1979.

CHAPTER TWO

THE EVOLUTION OF THE GENOME

NOTHING in biology has changed or is changing faster than our picture of the way the genetic material is organised in chromosomes. When I was a student, the chromosomes were seen as strings of beads; the beads, or genes, were thought to be proteins, and the string, nucleic acid, or DNA. Soon afterwards, the beads disappeared, and the DNA string carried the message; a chromosome became simply a set of DNA genes arranged end to end. By degrees, however, it emerged that there is an awful lot of DNA in the chromosomes of higher organisms, much of it present in large numbers of copies. Since genetic evidence shows that, typically, there are only two (or at most very few) genes of each kind in a cell, this was embarrassing. More recently, the techniques of nucleic acid sequencing and 'restriction mapping' (whereby the distribution of particular short sequences is discovered by cutting up the DNA with enzymes which recognise those sequences) have revealed a bewilderingly complex picture. At present, we have more facts than explanations. Therefore, a few specific topics have been chosen for discussion.

The papers in this section are likely to be difficult to follow, so I will first introduce some of the ideas and technical terms. A crucial process is that of recombination. In the familiar form of 'homologous recombination', two lengths of almost identical DNA (that is, two homologous chromosomes) break at identical points and rejoin after changing partners. The process is fundamental in evolution, because it provides a means whereby two favourable mutations arising in different ancestors can come together in a single descendent; biologically, the function of sex is to make recombination possible. However, recombination also enables pieces of DNA to move from one place in the genome to another, and is involved in the 'life cycles' of various elements (viruses, plasmids, etc.) which today live a parasitic existence inside cells, but which probably originated as parts of the genomes of higher organisms which got loose and set up on their own.

It is these elements which are the subject of Campbell's article, so a brief account of them follows. A 'bacteriophage' (or phage) is a virus which infects bacteria. Viruses have a genome which can be either DNA or RNA, and which multiplies inside a cell, using the cell's replicative machinery. The viral genome codes, among other things, for proteins which form a protective coat, enabling the virus to survive when the host cell dies and to infect new cells. The DNA of the much-studied 'temperate' bacteriophage, λ, can also be incorporated at a specific site into the genome of the bacterium, and is then replicated in time with the bacterium without killing it. Incorporation is brought about by recombination: both the replicating phage and the bacterial chromosome are circles, and recombination between two circles produces a single larger circle.

'Plasmids' are a second class of elements which differ from phages and viruses in that they do not code for coat proteins and so cannot survive outside a host cell; sensibly, they do not usually kill the cells they inhabit. They occur as independent replicating entities inside cells; some of them can also exist integrated into the chromosome. The remarkable F plasmids cause the bacterium they inhabit to 'conjugate' with another, thereby enabling the plasmid (as well as the bacterial chromosomes) to pass from one cell to another; a form of bacterial sex is caused by the presence of a parasite. A third class of elements are the 'transposons', which exist only as integrated parts of a chromosome, but which can multiply out of phase with it. Essentially, a transposon replicates, and the replica is incorporated into a new site on the chromosome, while leaving the original copy in place at the old site; hence there are two copies per genome where there was one before.

This leads us to the topic of 'selfish DNA'. The title derives from Dawkins' book, *The Selfish Gene*, which was original in proposing a gene's-eye view of life. Looked at from this point of view, it is clear that a transposon will tend to evolve characteristics which favour its own replication, whether or not those characteristics favour the survival of the cell it inhabits. This is the case argued in the papers by Doolittle and Sapienza, and by Crick and Orgel. The matter is of more general interest because some of the highly replicated elements in the chromosomes of higher organisms resemble in their structure the transposons of prokaryotes. It seems that our own chromosomes may be carrying around, interspersed between our 'own' genetic material, genes which are essentially parasitic.

The final paper in this section, by Jeffreys, is concerned with a different topic, the evolution of gene families, of which the haemoglobin genes are the best known. Some background information may help non-biologists to follow his paper. In adult vertebrates the haemoglobin molecule, which occurs in red blood cells and combines reversibly with oxygen, is composed of four polypeptides, two α chains and two β chains, the α and β chains being coded for by different genes. Evolutionary changes of the gene are of two kinds. 'Silent' changes alter a base in the DNA so that the resulting triplet of bases, or 'codon', still codes for the same amino acid; hence no change is caused in the resulting protein. 'Replacement' changes, in contrast, do cause a change in the protein. It is widely, but not universally, thought that silent changes are usually without effect on fitness—they are 'neutral'. As pointed out in 1968 by Kimura and by King and Jukes, neutral changes are likely to accumulate at a constant rate in evolution, thus providing a 'molecular clock'. Since that time it has become generally accepted that rapid evolutionary changes in molecular sequence is evidence that the sequence is selectively unimportant.

It has recently been discovered that most genes in higher organisms consist of sequences of bases, known as 'exons', which code for parts of the resulting protein, separated by intervening sequences, or 'introns', which are excised at the RNA stage and never translated into protein. There is much interest in the idea that the exons might once have been separate genes which have been brought together to code for a complex function.

The haemoglobin genes also illustrate the fact that, during evolution, a single gene may be duplicated, once or several times, with the duplicate copies diverging in structure and function. No doubt the α and β genes arose in that way. Today, mammals have a different haemoglobin in the foetus, consisting of two α and two γ chains; this makes it easier for oxygen to pass from mother to foetus. The gene for the γ chain arose by duplication of the β gene. Jeffreys reviews the number and arrangement of these genes in different animals, and speculates on their evolution. It turns out that there are also 'pseudogenes', which arose by duplication but have now lost their function.

Some general questions about movable elements and their implications

A. CAMPBELL

In this introductory chapter I present a highly selective history of research on movable genetic elements, followed by some comments on the molecular mechanisms of insertion and transposition and the possible roles of specific recombination systems in evolution and development. My general purpose is to call attention to some of the questions we should be thinking about and trying to answer.

History

TRANSPOSABLE ELEMENTS IN MAIZE

Barbara McClintock discovered movable elements in maize in the 1940s and proceeded to demonstrate a remarkable array of properties associated with them, such as controlled chromosome breakage, effects on gene expression, localized mutagenicity, etc.[1]. In many respects her work was far ahead of its time. It has been admired by geneticists because of its combined content of perception, biological insight, and experimental and analytical virtuosity, but the means to explore its full implications at a finer level are only now becoming available.

INSERTION

In the meantime, information obtained from bacterial genetics provided some relevant insights. The first of these was the realization that temperate bacteriophages such as λ and conjugative plasmids such as F can become physically inserted into the chromosome rather than joined to it by some looser type of connection. This idea seems so natural to us now that it is hard to imagine alternatives to it. However, I believe it is true that both in bacterial genetics through the 1950s and in the genetics of movable elements in higher eukaryotes during the same period, most investigators assumed that the mobility of the elements, their ability to add to chromosomes and to be lost from them, indicated that the connection between a movable element and the rest of the chromosome differed in its physical nature from the bonds that held the normal, 'immovable' parts of the chromosome together, such as the bonds connecting one gene with another.

Initially, the evidence for insertion came from fine-structure genetics, especially of bacteriophage λ, starting around 1960. Here advantage was taken of the enormous resolving power of bacterial and phage genetics, which permitted the demonstration that the genome of the inserted element is in fact colinear with that of the chromosome. More direct physical demonstration of insertion became possible with improvements in DNA technology in the 1970s. By 1972, Sharp *et al.*[2] had produced some remarkable electron micrographs of artificial three-way hetero-duplexes between the DNAs from λ viral particles, F plasmids, and an F' plasmid bearing an inserted λ prophage that were sufficiently definitive to convince even the most skeptical and unimaginative biochemist that insertion was a real process and that the participants were tangible objects amenable to experimental manipulation.

SITE-SPECIFIC RECOMBINATION

In the late 1960s another feature of λ insertion became evident, which has come to be called site-specific recombination. The underlying facts were, first, physiological evidence that λ insertion is controlled by the λ repressor and, later, the isolation by several investigators[3-5] of viral mutants that were unable to insert because of their failure to produce a specific gene product, λ integrase. Subsequently, Guarneros and Echols[6] and Kaiser and Masuda[7] showed that a second viral product, excisionase, was required for excision from the chromosome, but not for insertion.

The noteworthy implication here was that insertion and excision are processes in which the enzymes of homologous recombination, such as those encoded by the bacterial *rec* genes, play no part. Rather, they represent breakage and joining of DNA at specific sites by enzymes that recognize those sites and act on them alone.

Several more years elapsed before Nash[8] succeeded in developing a workable in vitro assay for the integrase reaction and before Landy and Ross[9] determined the nucleotide sequence of the junction points between prophage and host DNA. As a result of the combined efforts of many people, I think it is fair to say that the amount of direct knowledge about site-specific recombination in λ far exceeds that about any other element. Consequently, many of the generalizations that can be made at this time involve extrapolation from the λ work to other systems.

One such generalization is that most movable elements resemble λ in encoding some of the enzymatic machinery needed for the DNA-joining reactions in which they participate and that these element-encoded enzymes recognize specifically the termini of the element itself. This idea is supported by the evidence available on the relative handful of elements where this has been studied in depth. We may anticipate that this generalization is not absolute and depends more on the evolutionary pressures operative on the elements that have been studied to date than on any more basic principle.

TRANSPOSITION

The horizons of prokaryote geneticists were extended by the discovery[10-12] of inserted elements that can move about from one chromosomal location to another (like the McClintock factors) and the later expansion of this category to include elements that carry drug-resistance determinants and other genes with direct phenotypic effects. By the time attention became focused on these elements, molecular techniques were sufficiently advanced so that direct physical demonstration of insertion was simple, though not necessarily easy. This was a fortunate circumstance, because insertion sequences and transposons are inherently more difficult to study genetically than something like λ, which has a demonstrable free phase in which its genetics can be investigated and compared with that of the inserted state.

As these studies developed, some prokaryote geneticists, including myself, realized that we should have anticipated that elements of this sort ought to exist. The bacterial inserting elements that had been studied up to that time were temperate phages and plasmids that could replicate extrachromosomally and could also become inserted and replicated as part of the chromosome. Elements with this dual capacity were dubbed 'episomes' by Jacob and Wollman[13]. The course of prokaryotic work in the 1960s was strongly influenced by the desire for a unified concept of the interactions between episomes and chromosomes. Impressed by the similarity of episomes to the McClintock factors, at least some of us imagined that they would turn out to be episomes as well.

They still may. However, we already knew enough about the genetic organization of λ more than 10 years ago to appreciate that the ability to replicate autonomously and the ability to insert are independent attributes. Many phages, including mutants of λ, can replicate but cannot insert. Someone might have thought more seriously about the possible existence of natural elements capable of insertion but not of independent replication and interpreted existing data in that light.

The story has an ironic twist. As the analogy between transferable prokaryotic elements and temperate phages like λ was gradually coming to seem natural, comfortable, and familiar, evidence was developing that the mechanism of transposition is, in fact, fundamentally different from the excision-insertion pathway known in λ. In λ infection, viral DNA enters the cell and assumes a closed circular form. Insertion comprises a single reciprocal exchange taking place somewhere within a 15-bp segment of the viral DNA and an identical 15-bp segment that preexists at a unique site on the chromosome. Excision of prophage from the chromosome reverses this process. No DNA is degraded, and there is no new synthesis. The reaction is remarkable in that its representation as a precise reciprocal exchange has continued to hold true through successively finer levels of analysis both in vitro and in vivo.

Transposition of an insertion sequence like IS1 as observed in vivo is quite different. Like λ, the inserted element is flanked by a direct oligonucleotide repeat, in this case 9 bp long. Translocation to new sites comprises the appearance of the element at the new site without concomitant loss from the old site. Obviously, DNA synthesis must accompany transposition. Not only is the element itself duplicated, but also 9 bp of DNA originally in the recipient, the two copies of which now flank the insert at its new site. Hence, not only does the basic process differ from λ insertion, but in these elements replication and insertion are connected rather than independent.

DNA REARRANGEMENTS

The last topic in this brief history is the discovery of special systems in which DNA rearrangements play a demonstrable role in controlling gene expression or cellular differentiation. These include the control of flagellar antigen synthesis in *Salmonella* phase variation, the determination of mating type in yeast, and the differentiation of vertebrate lymphocytes to form specific types of immunoglobulins. The existence of these systems has focused attention on the possible importance of controlled DNA rearrangements throughout the biological world.

Mechanism

In considering the mechanism of site-specific recombination, if we restrict ourselves to systems in which a fairly well-defined reaction can be studied in vitro, there is only one system available, λ integrase. Rather than review the biochemistry of this one system, I would like, instead, to explore the implications of the foregoing discussion that λ insertion and IS1 transposition represent basically different processes. One of the tasks of scientists is to search for syntheses that illuminate whatever unity may be discernible beneath diversity. The time for such a synthesis may be premature. However, it is appropriate to focus our attention on what features these two types of reactions may have in common and whether the enzymes carrying them out might be derived by minor modifications from a common progenitor.

First, what do the two have in common? One feature, which to my knowledge is shared by all prokaryotic insertion systems, is that the inserted element ends up flanked by a direct repeat—frequently of oligonucleotide length but sometimes, as in the case of the F factor, much longer. The direct repeats appear to be there for different reasons in various cases.

With elements like λ, which insert preferentially at a unique site or at a small number of sites, the flanking repeat is present in the element itself and appears to be part of the DNA recognition site for the enzyme. On the other hand, transposable elements like IS1 insert at many sites, and there is apparently no specific recognition of the recipient DNA oligonucleotide sequence that ends up flanking the inserted element. What is generally found with such elements is an imperfect inverted repeat at the ends of the inserted elements, as though the two ends had evolved so as to look equivalent to some recognition system when viewed from either end.

Table 1 shows a few systems, mostly from prokaryotes, that involve site-specific recombination. These can be classified into two groups depending on whether they resemble λ insertion or IS1 transposition. I have labeled such systems 'conservative' and 'duplicative,' respectively. A conservative reaction is a simple polynucleotide exchange. Nothing is created, nothing is destroyed; DNA is simply rearranged. In a duplicative reaction, on the other hand, there is a net increase in DNA.

Table 1   Site-specific Recombination Systems

Conservative reactions	Duplicative reactions
Insertion of λ, P2, P22, etc.	IS and Tn transpositions
Excision of λ, etc.	Mu insertion
Specific inversions:	Replicon fusion
Salmonella phase variation	Retrovirus insertion?
G loop of Mu	
Clean excision of IS and Tn	
elements	

In a very strict sense, the only demonstrated conservative reactions are λ insertion and excision, where the fate of substrate molecules can be followed directly in vitro. Nevertheless, there is a strong presumption that some of the other systems listed are conservative.

Of special interest are the controlled inversions, such as the G segment of Mu and the segment that regulates phase variation in Salmonella. Inversion, like λ insertion, is a simple rearrangement in which the net DNA content is unchanged. We can imagine a controlled inversion system evolving directly and immediately from a λ prophage. Synthetic invertible elements can, in fact, be derived from λ by appropriate manipulation[14]. The Salmonella phase variation system resembles such a λ model at least insofar as the termini of the invertible segment are two identical oligonucleotide sequences in opposite orientation.

The duplicative reactions include transposition by all of the common insertion sequences and transposons of the Enterobacterioceae and also bacteriophage Mu, which utilizes the transposition mechanism not only for inserting into the chromosome and for transposing from one site to another, but also in its own autonomous replication. I have also tentatively placed retroviruses in this category.

To get back to the basic question: Are there any similarities in the two types of processes that might encourage one to look for common aspects of mechanism? If not, is there any way to seek relevant evidence?

To my mind, the most direct evidence that some steps are common to the two processes would be if someone found one element capable of carrying out both types of processes and could show that the same specific gene and/or its enzymatic product is required for both. There are some suggestive cases with known elements, but I am unaware of any for which there is really hard evidence. To put the question in terms of synthetic invertible elements derived from λ, it is clear that such an element is expected to invert at high frequency in the presence of integrase. I would not expect it to transpose in the presence of integrase, but my mind is open to the possibility that it might do so at low frequency.

The general idea is illustrated in Figure 1. The figure represents successive steps in concerted reactions, not independent reactions with free intermediates. The donor is our artificial λ with inverted ends. The scheme has been drawn to incorporate a 'sticky-ended' intermediate which has often been postulated. For our purposes, such an intermediate provides a convenient, but not necessarily unique, mechanism for promoting homologous recombination within a short DNA segment. We imagine that formation of this intermediate is catalyzed by a sequence-specific, element-encoded enzyme, perhaps acting in conjunction with other enzymes. This first step is then represented as leading to either inversion or transposition depending on what happens next.

In the inversion reaction, the element is reinserted, in opposite orientation, in the donor chromosome. On the transposition side, a recipient chromosome is cleaved by a host enzyme that makes staggered nicks, which must generate ends of the opposite chemical polarity from the original reaction; that is, if the original cuts generated 5' ends, these must be 3'. The projecting single-stranded ends are then joined, and repair synthesis fills in the gaps.

The purpose of this diagram is to indicate how a single enzyme, or two related variants of the same enzyme, might participate in both types of reactions, depending in part on its association with additional host enzymes. Now I must return to the fact that to simplify things I have cheated quite a lot on the transposition side. Among other things, I have diagrammed a mechanism for transposition but not one that is duplicative. I have, in fact, destroyed the donor chromosome by cleaving it in two.

What can become of it? We know of no mechanism whereby it can rejoin and heal, because its sticky ends are identical rather than complementary. The easiest way to rejoin it is to reinsert the excised DNA. But to both reinsert and transpose, that DNA must be duplicated. How? It could replicate as such, with some special attention to what happens to the ends, to generate one transposable copy and one reinsertable copy.

Somewhat more realistic schemes, some elements of which are incorporated in Figure 1, have been proposed by others[15,16]. Their models invoke a host-enzyme-induced cleavage of recipient DNA to produce sticky ends, which are then ligated to element DNA. Unlike Figure 1, these models do not require production of sticky ends at the ends of the insert. Rather, a cleavage of one of the two strands at each end suffices. (The larger arrows of Fig. 1 represent the cleavage sites in these models.)

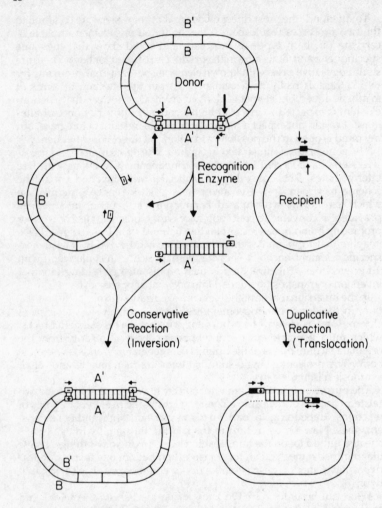

FIG. 1. *Hypothetical scheme illustrating one possible manner in which a single enzyme might participate in both conservative and duplicative reactions. The donor molecule contains terminal oligonucleotide repeats which, like the 15-bp common core of λ, function both as recognition elements and as sites of action, but which, unlike the λ case, are in inverted orientation. An element-encoded 'recognition enzyme' (like λ integrase) makes staggered nicks within the common core regions at both ends (not necessarily simultaneously). The excised molecule then has one of two possible fates: (1) It may, as shown on the left, reinsert into the donor molecule in inverted orientation. This is the standard integrase reaction as applied to a substrate with inverted ends. (2) It may transpose by inserting into a recipient molecule that has been cleaved in the appropriate manner by host enzymes, followed by repair synthesis to generate oligonucleotide repeats. More detailed discussion of the duplicative reaction is given in the text.*

This is of course simpler than excising the insert, replicating it, and then reinserting a copy. In these schemes, the insert DNA never leaves its original location and hence does not need to be reinserted. There is an additional simplification in that the replication of the insert and the gap-filling that I depicted become part of one replication event rather than two separate ones. Shapiro's scheme postulates, in addition to the enzyme active at the terminus, a second specific enzyme whose site of action is internal to the element and whose mode of action is conservative rather than duplicative.

What do schemes like this predict with respect to the prospects for success in looking for one enzyme that participates in both conservative and duplicative reactions? Even from Figure 1 it is obvious that the recognition enzyme must act in concert with other enzymes and that a given enzyme may therefore be specialized to function in one manner or the other. In that case, Figure 1 may represent not so much the potentialities of a single enzyme as the collective potentialities of a family of closely related enzymes, perhaps with overlapping specificities. Carried a step further, an enzyme that makes two nicks rather than four, as in the Grindley and Shapiro models, would also represent a variant of a family that includes conservative enzymes as well. These considerations encourage a comparative approach and justify skepticism about the likelihood that a single enzyme might function in both manners.

Two further comments on natural transposons and insertion sequences: The termini of the element generally form an imperfect inverted repeat, not a perfect one. This suggests that they are not frequently the sites of conservative reactions, which should depend on precise homology. Second, precise excision does occur (Table 1), although it does not seem to represent a step in transposition. Precise excision, with restoration of gene function, presumably requires recombination between the oligonucleotide repeats of host DNA that flank the element. The reciprocal product is generally not recoverable; so characterization of the reaction as conservative is tentative.

At past Symposia the introductory remarks have sometimes included predictions. It is against my principles to make predictions about science. However, I enjoy the more harmless recreation of making predictions about scientists. I predict that at some stage of the game scientists will dedicate serious effort to understanding the relation between conservative and duplicative reactions. This may happen because the data indicate that such efforts will be auspicious; but even if they are not, the question will be examined in depth because it offers our only hope for a unified picture. Such studies may concentrate either on single totipotent enzymes or on families of recognizably related enzymes.

Evolution

When Barbara McClintock discovered controlling elements in the 1940s, it was obvious that they might play important roles both in evolution and in development. In presenting her results she generally concentrated more attention on the developmental implications than on the evolutionary ones. I believe that this was a wise decision on her part, although I do not know whether her actual reasons for this were the same as mine.

Changes in DNA, ranging from point mutations through gross rearrangements, are the raw material of evolutionary change. We cannot doubt that any mechanism that produces such changes must have occasionally contributed to evolution by creating adaptive gene combinations that have survived. But it is also true that at the time controlling elements were discovered, there was no obvious shortage of known mechanisms (understood at the same level as controlling elements could be understood at the time) that could generate the same end results. Evolutionists did not accord high priority to the search for new mutational mechanisms. The principal outstanding problems in their field did not appear to be at that level.

Although some people may disagree, I believe that, for the most part, the situation remains the same today. We all agree that movable elements can cause deletions, inversions, translocations, etc. Few of us would assert that all deletions or all inversions require their participation. With a few possible exceptions, such as the distribution of dispersed, moderately repetitive DNA within related species, most observable evolutionary change can be adequately (though not necessarily correctly) explained by other mechanisms. Without in any way minimizing the value of some interesting evolutionary speculations that have been made recently, I believe that the case that movable elements provide much of unique explanatory value remains to be made.

One example often cited of the role of specific events in evolution is the bacterial plasmids. Comparisons of the genomic organization of related plasmids show that frequently they differ from one another by insertion of transposons and by replicon fusions catalyzable by specific systems. There can be no reasonable doubt that these events have occurred in nature and have contributed to the diversity of the existing plasmid population.

Accepting that such events have happened and are important, one may still ask whether they really constitute evolution. Let me illustrate the reasoning behind this position with an extreme case. If I survey *Escherichia coli* strains in nature, some are lysogenic for bacteriophage λ and many are not. I could, therefore, speak of the acquisition of a λ as part of the evolution of a strain such as *E. coli* K12. In a sense it is. However, perhaps because of conditioning, I generally regard the transition from phage to prophage and back again as defining the life cycle of the virus, rather than as constituting evolution, which implies (to me) some degree of progressive as well as cyclical change. I tend to consider the relocation of transposons and the insertion of one plasmid into another likewise as constituting part of the life cycle of these elements. Unions are created when convenient and eventually dissolved when their selective value disappears. All this happens within a single bacterial species or a group of species belonging to some Exchanger List, and I suspect it happens at a frequency many times as great as events that lead to a degree of genetic isolation comparable to that separating distinct species of higher organisms.

This is not the time to make a judgment concerning this issue. It may be profitable for bacterial geneticists to consider carefully where they want to draw the line between evolutionary change and variation that is part of the ongoing population biology of the species. They may wish to relegate cyclical processes and pedigrees that are reticulate rather than branching to the latter category. Perhaps, on the other hand, it is impracticable to make the distinction between evolutionary change and nonevolutionary change and we should simply chronicle variation as such. Even in that event, I would prefer to see that decision made consciously and explicitly rather than, as it seems at the moment, implicitly and by default.

Development

With those remarks, I shall leave evolution and turn to the question that is really the major thrust of the current interest in movable elements, namely, their potentialities for controlling gene action, especially during development.

If evolutionists have generally perceived an abundance of possible mechanisms for generating genetic diversity during evolution, genetically oriented developmentalists contemplating the differentiation of the descendants of a single zygote into more or less permanently distinct types have never enjoyed that luxury. They have, in fact, generally been hard pressed to find any precedents or paradigms in the transmission genetics of organisms, including unicellular ones, that explain the changes in cellular properties during development in a manner that is either satisfactory or satisfying. The classical notion of differentiation as a process that generates changes that are permanent and heritable at the cellular level does not square with the textbook picture of equal distribution of genes between daughter cells whenever a somatic cell divides by mitosis. The demonstration by McClintock that movable elements could translocate to positions adjacent to known genes and permanently modify their expression and that translocation is subject to some degree of temporal control provided a genetic system with many of the features needed to understand how changes could be at once inducible and permanent.

A role of controlling elements in developmental specification implies that the critical relevant changes are nuclear rather than cytoplasmic. On this important question, data from nuclear transplantation experiments have been equivocal. They have sometimes been interpreted as indicating that nuclei from adult cells retain the totipotency of the zygote nucleus, but that conclusion can be questioned on a technical level[17]. I have always found the conclusion conceptually unsatisfying as well, because it seems to beg the question of what is the basis of permanent change, if not genetic.

In any event, there is a growing awareness of the possibility that controlled DNA arrangements are critical in development, and there are a few supporting facts. The most impressive are from antibody-forming cells. Extrapolation from that system to general developmental mechanisms requires some optimism, because part of the motivation for studying immunoglobulin determination is that even among developmental systems, its properties are extraordinary.

There are examples from prokaryotes of the recombinational control of gene expression whose mechanisms are better understood. Paramount among these are the controlled inversions, such as the G region of Mu or the H-antigen control region of *Salmonella*. A gene outside an invertible region transcribed from a promoter inside the region is turned on when the inversion is in one orientation (flip); in the other orientation (flop), the gene is off. Developmental biologists, facing the question of why a given gene is on in one tissue and off in another, though present in both, may well be attracted by the potentialities of such a picture.

I think that most integrative biologists would also agree that, viewed as a model for developmental specification, the system lacks one important feature. It does show us how a specific recombination system can control expression of a gene. That is clearly one aspect of development. It does not include or imply a mechanism by which the recombination itself is controlled so as to fit into the overall developmental program of the organism. Such a program must embody not only the capacity to change, but also the ability to change in a highly directed manner, so that the right switch can be thrown in a prescribed cell at a prescribed time in development. It is in this regard that immunoglobulin synthesis may be atypical, because the system is designed to generate a high level of somewhat random diversity, whereas the typical critical changes in development should be much more tightly controlled.

If we consider the prokaryotic model systems, we can ask whether they are under tight control. The recombinational event in the *Salmonella* switch is controlled by the product of a specific gene within the element. That gene may be subject to some quantitative effects on its rate of action, but there is no evidence that it is tightly regulated. Furthermore, there is no reason to suppose that it should be. The natural role of phase variation seems to be to counter the immunological defenses of the host by creating a more rapid rate of antigenic change than is readily accomplished by mutation alone. That purpose is satisfied by a reversible reaction that occurs constitutively. In development, on the other hand, the critical changes should be regulated and irreversible.

Obviously, there are many mechanisms imaginable whereby site-specific recombination could be controlled. Prokaryotic systems provide some interesting examples of how it is controlled in those systems. I would like to describe some aspects of the regulation of λ insertion and excision in that context.

There is a special reason (other than my own involvement in the work) for talking about λ rather than simpler elements like Tn3 or Tn5. λ represents a genetic element that is sufficiently complex so that it has a regulatory program that includes genes for repression, replication, assembly, and lysis, as well as insertion. The insertion genes must therefore be controlled so as to fit into this program, just as we may expect that analogous developmental events are controlled by the program of the whole organism. It is in this respect that I regard λ regulation as deserving the attention of biologists in general.

As with other aspects of λ biology, the control of site-specific re-combination has many facets and has achieved a level of sophistication that may well be unmatched in organisms like higher eukaryotes, which are larger and genetically more sluggish. The whole story has been summarized by Miller[18]. Here I shall discuss only one aspect, namely, the differential control of *int* and *xis* transcription, and only enough of that to indicate some critical features of the story.

For this purpose it may be helpful to look at λ in a particular manner. λ is a virus with a genome of about 50,000 nucleotide pairs. Of this DNA, all of the functions and sites concerned with integration are clustered into one segment about 1500 nucleotides in length. I like to imagine that a virus like λ evolved by the joining together of different modules that previously functioned in isolation or in other contexts. According to that view, these 1500 bp might be derived from and phylogenetically related to simpler insertion elements like IS1 or IS2.

This viewpoint has two relevant consequences. The first relates to the discussion of the relationship between conservative and duplicative re-actions. Modular concepts of evolution have become popular in the last few years mainly because the properties of transposons suggested a specific mechanism whereby modules might come together. I find that most of my colleagues are fairly well conditioned to accepting the notion that the insertion region of λ might have a common evolutionary origin with some element like IS2. Fewer people seem willing to follow through with what I consider a reasonable corollary, namely, that the common ancestry of the two types should be reflected, at some level, in a commonality of mechanism.

The second feature is the one I wish to address now. Compared with an IS element, λ is a large and complicated entity with a well-ordered life cycle. Hence, the potentialities of the insertion region must at some stage have come under the control of the λ program. How is this accomplished?

Obviously, the relationship between λ and its insertion genes is not the same as that between an organism and a movable element within its genome. The insertion region of λ functions to insert the whole λ genome within the bacterial genome, rather than moving about within the λ genome. Let us ignore that detail for the moment and focus on the regulatory problem common to both cases of subordinating the re-combination functions to the program of the whole organism.

The basic problem concerning regulation of λ insertion was already perceptible 10 years ago. Insertion of λ into the *E. coli* chromosome requires only integrase. Excision from the chromosome requires another viral protein, excisionase, as well. The *int* and *xis* genes lie next to each other on the λ DNA and are both included in the major leftward tran-script of the virus.

Figure 2 illustrates the lifestyle of λ and its attendant problems. When cells are infected with λ, a fraction of the cell population goes into the productive cycle, eventually dying and liberating virus. In such cells, insertion of viral DNA into the chromosome is a waste, and insertion without excision may cause the infection to abort. Hence, one would like

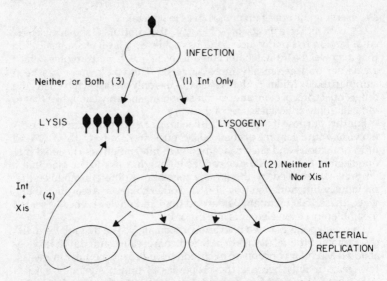

FIG. 2. *Life cycle of* λ, *showing that the need for three positions of the regulatory switch controlling integrase production: (1) int* on, *xis* off *(in those infected cells destined to survive as lysogens); (2) both* off *(in established lysogenic bacteria); (3) both* on *(in induced lysogenic bacteria, and perhaps in infected cells destined for lysis).*

to see *int* and *xis* either both turned on or both turned off. Certainly, integrase alone would be bad. Another fraction of the cell population survives infection because viral functions become repressed. It is in this fraction that insertion, without subsequent excision, is desired. If a surviving cell harbors a repressed, unintegrated phage genome, that genome is eventually diluted out with growth. So here we want integrase alone.

After insertion has taken place and the stable lysogenic condition is established, neither integrase nor excisionase is needed to perpetuate that condition. Integrase alone would not hurt anything; integrase and excisionase together would cause prophage loss. Finally, within the lysogenic population occasional cells become derepressed and go over into the productive cycle. In this case, integrase and excisionase are needed to allow those cells to produce phage.

The mechanism whereby these ends are achieved was worked out between 1970 and 1980 by various people in several laboratories. Figure 3 summarizes the results. The *int* gene can be transcribed from two different promoters—P_L and P_I. P_L is the major leftward promotor. Transcription from P_L is observed after infection and following derepression of a lysogenic cell. In that transcript *int* and *xis* are expressed coordinately. This assures that excision will follow induction and that productively infected cells will not lose viral genomes by burying them in the chromosome. Transcripion from P_I, on the other hand, leads to expression only

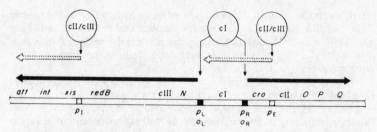

FIG. 3. *Structure of three-way switch controlling integrase and excisionase production in* λ. *Position 1 (int on, xis off) is achieved by turn on of transcription from P_I, which produces a message translated to give integrase but not excisionase. This message is turned on by the cII/cIII proteins, which simultaneously turn on the cI gene. The cI product (repressor) turns off transcription from P_L and P_R, allowing cell survival and also shutting off excisionase production from P_I. Position 2 (both off) is reached when the cI product is elaborated continuously in a lysogenic bacterium. Repressor directly shuts off int and xis transcription from P_L. It also prevents transcription from P_I, by blocking transcription of the cII and cIII genes. In position 3, the P_L message is made, resulting in coordinate expression of int and* xis *without sufficient accumulation of cII/cIII to turn on P_I and P_E.*

of *int*. The P_I promotor is turned on by the products of genes *c*II and *c*III. These two proteins are also required to turn on the transcription of the *c*I gene, which codes for repressor. Therefore, in those cells that build up a high concentration of *c*II and *c*III, repression becomes established, and it is in those same cells that high levels of integrase are induced. Thus, the two aspects of lysogenization, repression and insertion, are coordinated, not because the two genes are in the same operon, but because two different operons respond to the same control system.

Finally, in the established lysogenic cell, repressor turns off both the transcription of *int* and *xis* from P_L and also the transcription of *c*II and *c*III, which are needed to activate P_I. Thus, both pathways are off, and no energy is wasted on useless synthesis.

Detailed molecular mapping places the P_I promoter, not between the *int* and *xis* genes, but straddling the beginning of the *xis* gene[19]. The initiation codon of the *xis* gene is within the Pribnow box of this promoter, so that almost the entire *xis* gene is transcribed as a leader for the *int* message. This location of the promoter is intriguing with respect to the evolutionary considerations mentioned earlier. If the insertion region of λ is descended from a free-living IS element, it might seem reasonable that the regulated promoters putting the genes of the element under λ control lie outside the region itself, or at the border between the insertion region and the rest of λ.

With this incomplete and simplified description of λ regulation as an example, I would like now to return to some more general questions concerning developmental specification and differentiation.

If the critical switching event that commits a cell to a particular pathway is a change in DNA involving systems like those known in prokaryotes, at least two requirements must be met. First, the relevant recombinations must, as in the λ example, be regulated so as to function at the proper time. Second, there must be some mechanism that makes the switching event effectively (though not necessarily absolutely) irreversible. A process that allows free switching back and forth, even during a brief window in time, would not do.

Table 2 lists several possible methods for achieving irreversibility. Duplicative reactions, by their nature, are not directly reversible. The basic mechanism creates a new element without destroying an old one. Iteration of that operation will not restore the original condition. Excision events, which apparently occur in immunoglobulin synthesis, are irreversible even if the underlying chemical process is both conservative and symmetrical. This is because the reciprocal products suffer different fates depending on whether or not they contain a centromere (in eukaryotes) or an active replication origin (in prokaryotes).

Table 2   Mechanisms for Irreversibility of Site-specific Recombination

Mechanism	Example
Duplicative reaction	transposition of IS1
Excision	loss of IS1
Site recognition (*int-xis* type)	insertion of λ prophage
Site regulation of enzymes	insertion of P2 prophage

Differential recognition of the sites involved in the forward and reverse reactions, as exhibited by λ, is another mechanism. The basis for the different catalytic specificities of insertion and excision lies in the differential recognition of the base sequences flanking the phage and bacterial sites. Thus, in the presence of integrase alone the reaction is effectively irreversible.

Finally, the rearrangement event itself may influence the expression of the genes whose products generate the rearrangement. One example of such recombinational control of recombination functions is seen in bacteriophage P2, where the insertion site lies within the *int* operon, which is therefore inactivated by the insertion event[20]. That effect, without differential site recognition, would suffice to render insertion irreversible. In bacteriophage λ, also, there is good evidence that *int* expression is affected by downstream DNA that is contiguous to it in the phage, but not the prophage, state, although the nature of the effect in λ is somewhat different from that in P2[18].

With respect to developmental programs, the last two mechanisms, especially, can be described with a computer analogy. We may imagine a specific cell receiving an instruction of the form, 'Now is the time to move from square one to square two.' We may, more realistically, dissect this instruction into two components: External information, both temporal and positional, impinges on the cell and is read as, 'If you are now in square one, move to square two.' Execution of the order requires internal, propioceptive information that answers the question, 'Which square am I in right now?' The last two mechanisms also offer the possible advantage of being effectively irreversible under certain circumstances and potentially reversible under others.

The organizers of this meeting expressed the hope that 'the Symposium will focus on the mechanisms by which specific genetic rearrangements occur and the mechanisms by which these rearrangements may control gene expression.' We may predict that if and when another Symposium is held covering some of the same ground as this one, the focus will be on a third question as well: It will focus not only on how rearrangements occur and how rearrangements control gene expression, but also on the mechanisms by which the occurrence of specific rearrangements is itself controlled and integrated into the overall biology of the organism. At that point molecular biologists may fairly claim to have come to terms with the major implications of McClintock's classical work, in which these same questions were posed at a different level.

1. McClintock, B. Chromosome organization and genic expression. *Cold Spring Harbor Symp. Quant. Biol.* **16**, 13 (1952).
2. Sharp, P., Hsu, M. & Davidson, N. Note on the structure of prophage λ. *J. Mol. Biol.* **71**, 499 (1972).
3. Zissler, J. Integration-negative (*int*) mutants of phage λ. *Virology* **31**, 189 (1967).
4. Gingery, R. & Echols, H. Mutants of bacteriophage λ unable to integrate into the host chromosome. *Proc. Natl. Acad. Sci.* **58**, 1507 (1967).
5. Gottesman, M. & Yarmolinsky, M. Integration negative mutants of bacteriophage lambda. *J. Mol. Biol.* **31**, 487 (1968).
6. Guarneros, G. & Echols, H. New mutants of bacteriophage λ with a specific defect in excision from the host chromosome. *J. Mol. Biol.* **47**, 565 (1970).
7. Kaiser, A. & Masuda, T. Evidence for a prophage excision gene in λ. *J. Mol. Biol.* **47**, 557 (1970).
8. Nash, H. A. Integrative recombination of bacteriophage lambda DNA *in vitro. Proc. Natl. Acad. Sci.* **72**, 1072 (1975).
9. Landy, A. & Ross, W. Viral integration and excision: Structure of the lambda *att* sites. *Science* **197**, 1147 (1977).
10. Jordan, E., Saedler, H. & Starlinger, P. O^c and strong polar mutations in the *gal* operon are insertions. *Mol. Gen. Genet.* **102**, 353 (1968).
11. Shapiro, J. A. Mutations caused by the insertion of genetic material into the galactose operon of *Escherichia coli. J. Mol. Biol.* **40**, 93 (1969).
12. Fiandt, M., Szybalski, W. & Malamy, M. H. Polar mutations in *lac, gal* and phage λ consist of a few DNA sequences inserted in either orientation. *Mol. Gen. Genet.* **119**, 223 (1972).
13. Jacob, F. & Wollman, E. Les épisomes, élements génétique ajoutes. *C. R. Acad. Sci.* **247**, 154 (1958).
14. Reyes, O., Gottesman, M. & Adhya, S. Formation of lambda lysogens by IS2 recombination: *gal* operon-lambda P_R promoter fusion. *Virology* **94**, 400 (1979).
15. Grindley, N. D. F. & Sherratt, D. J. Sequence analysis at IS1 insertion sites: Models for transposition. *Cold Spring Harbor Symp. Quant. Biol.* **43**, 1257 (1979).
16. Shapiro, J. A. Molecular model for the transposition and replication of bacteriophage Mu and other transposable elements. *Proc. Natl. Acad. Sci.* **76**, 1933 (1979).
17. McKinnell, R. G. *Cloning—Nuclear transplantation in amphibia.* University of Minnesota Press, Minneapolis (1978).
18. Miller, H. *Cold Spring Harbor Symp. Quant. Biol.* XLV, 186 (1981).
19. Abraham, J., Mascarenhas, D., Fischer, R., Benedik, M., Campbell, A. & Echols, H. DNA sequence of regulatory region for integration gene of bacteriophage λ. *Proc. Natl. Acad. Sci.* **77**, 2477 (1980).
20. Bertani, L. Split operon control of a prophage gene. *Proc. Natl. Acad. Sci.* **65**, 331 (1970).

A. Campbell is in the Department of Biological Sciences, Stanford University.

This article first appeared in *Cold Spring Harbor Symposia on Quantitative Biology* Vol. XLV, pp. 1–9; 1981. It is reproduced with permission.

Selfish genes, the phenotype paradigm and genome evolution

W. FORD DOOLITTLE AND CARMEN SAPIENZA

Natural selection operating within genomes will inevitably result in the appearance of DNAs with no phenotypic expression whose only 'function' is survival within genomes. Prokaryotic transposable elements and eukaryotic middle-repetitive sequences can be seen as such DNAs, and thus no phenotypic or evolutionary function need be assigned to them.

THE assertion that organisms are simply DNA's way of producing more DNA has been made so often that it is hard to remember who made it first. Certainly, Dawkins has provided the most forceful and un-compromising recent statement of this position, as well as of the position that it is the gene, and not the individual or the population, upon which natural selection acts[1]. Although we may thus view genes and DNA as essentially 'selfish', most of us are, nevertheless, wedded to what we will call here the 'phenotype paradigm'—the notion that the major and perhaps only way in which a gene can ensure its own perpetuation is by ensuring the perpetuation of the organism it inhabits. Even genes such as the segregation-distorter locus of *Drosophila*[2], 'hitch-hiking' mutator genes in *Escherichia coli*[3,4] and genes for parthenogenetic reproduction in many species[4]—which are so 'selfish' as to promote their own spread through a population at the ultimate expense of the evolutionary fitness of that population—are seen to operate through phenotype.

The phenotype paradigm underlies attempts to explain genome structure. There is a hierarchy of types of explanations we use in efforts to rationalize, in neo-darwinian terms, DNA sequences which do not code for protein. Untranslated messenger RNA sequences which precede, follow or interrupt protein-coding sequences are often assigned a pheno-typic role in regulating messenger RNA maturation, transport or transla-tion[5-7]. Portions of transcripts discarded in processing are considered to be required for processing[8]. Non-transcribed DNA, and in particular repetitive sequences, are thought of as regulatory or somehow essential to chromosome structure or pairing[9-11]. When all attempts to assign a given sequence or class of DNA functions of immediate phenotypic benefit to the organism fail, we resort to evolutionary explanations. The

DNA is there because it facilitates genetic rearrangements which increase evolutionary versatility (and hence long-term phenotypic benefit)[12-17], or because it is a repository from which new functional sequences can be recruited[18,19] or, at worst, because it is the yet-to-be eliminated by-product of past chromosomal rearrangements of evolutionary significance[9,19].

Such interpretations of DNA structure are very often demonstrably correct; molecular biology would not otherwise be so fruitful. However, the phenotype paradigm is almost tautological; natural selection operates on DNA through organismal phenotype, so DNA structure must be of immediate or long-term (evolutionary) phenotypic benefit, even when we cannot show how. As Gould and Lewontin note, 'the rejection of one adaptive story usually leads to its replacement by another, rather than to a suspicion that a different kind of explanation might be required. Since the range of adaptive stories is as wide as our minds are fertile, new stories can always be postulated' (ref. 20).

Non-phenotypic selection

What we propose here is that there are classes of DNA for which a 'different kind of explanation' may well be required. Natural selection does not operate on DNA only through organismal phenotype. Cells themselves are environments in which DNA sequences can replicate, mutate and so evolve[21]. Although DNA sequences which contribute to organismal phenotypic fitness or evolutionary adaptability indirectly increase their own chances of preservation, and may be maintained by classical phenotypic selection, the only selection pressure which DNAs experience directly is the pressure to survive within cells. If there are ways in which mutation can increase the probability of survival within cells without effect on organismal phenotype, then sequences whose only 'function' is self-preservation will inevitably arise and be maintained by what we call 'non-phenotypic selection'. Furthermore, if it can be shown that a given gene (region of DNA) or class of genes (regions) has evolved a strategy which increases its probability of survival within cells, then no additional (phenotypic) explanation for its origin or continued existence is required.

This proposal is not altogether new; Dawkins[1], Crick[6] and Bodmer[22] have briefly alluded to it.

However, there has been no systematic attempt to describe elements of prokaryotic and eukaryotic genomes as products of non-phenotypic selection whose primary and often only function is self-preservation.

Transposable elements in prokaryotes as selfish DNA

Insertion sequences and transposons can in general be inserted into a large number of chromosomal (or plasmid) sites, can be excised precisely or imprecisely and can engender deletions or inversions in neighbouring chromosomal (or plasmid) DNAs[12-16]. These behaviours and, at least in some cases, the genetic information for the enzymatic machinery involved, must be inherent in the primary sequences of the transposable elements themselves, which are usually tightly conserved[12-16,23]. Most speculations on the function of transposable elements concentrate on the role these may have, through chromosomal rearrangements and the modular assembly of different functional units, in promoting the evolution of plasmid and bacterial chromosomes, and thus in promoting long-term phenotypic fitness[12-16]. Most assume that it is for just such functions that natural selection has fashioned these unusual nucleic acid sequences.

Although transposable elements may well be beneficially involved in prokaryotic evolution, there are two reasons to doubt that they arose or are maintained by selection pressures for such evolutionary functions.

First, DNAs without immediate phenotypic benefit are of no immediate selective advantage to their possessor. Excess DNA should represent an energetic burden[24,25], and some of the activities of transposable elements are frankly destructive[12-16]. Evolution is not anticipatory; structures do not evolve because they might later prove useful. The selective advantage represented by evolutionary adaptability seems far too remote to ensure the maintenance, let alone to direct the formation, of the DNA sequences and/or enzymatic machinery involved. A formally identical theoretical difficulty plagues our understanding of the origin of sexual reproduction, even though this process may now clearly be evolutionarily advantageous[1,4].

Second, transposability itself ensures the survival of the transposed element, regardless of effects on organismal phenotype or evolutionary adaptability (unless these are sufficiently negative). Thus, no other explanation for the origin and maintenance of transposable elements is necessary. A single copy of a DNA sequence of no phenotypic benefit to the host risks deletion, but a sequence which spawns copies of itself elsewhere in the genome can only be eradicated by simultaneous multiple deletions. Simple translocation (removal from one site and insertion into another) does not provide such insurance against deletion. It is significant that recent models for transposition require retention of the parental sequence copy[26,27], and that bacterial insertion sequences are characteristically present in several copies per genome[16]. The assumption that transposable elements are maintained by selection acting on the cell does not require that they show these characteristics. The evolutionary behaviour of individual copies of transposable elements within the environment represented by a bacterial genome and its descendants can be understood in the same terms as organismal evolution. Replicate copies of a given element may diverge in sequence, but at least those features of sequence required for transposition will be maintained by (non-phenotypic) selection; copies which can no longer be translocated will eventually suffer elimination. Some divergent copies may be more readily transposed; these will increase in frequency at the expense of others. Transposable elements which depend on host functions run the risk that host mutants will no longer transpose them; it is significant that at least some transposition-specific functions are known to be coded for by the transposable elements themselves[26-29]. It is not to the advantage of a transposable element coding for such functions to promote the transposition of unrelated elements; the fact that given transposable elements generate flanking repeats[16,30] of chromosomal DNAs of sizes characteristic to them (that is, 5, 9 or 11–12 base pairs) may indicate such a specificity in transposition mechanism. It is to the advantage of any transposable element to acquire genes which allow independent replication (to become a plasmid), promote host mating (to become a self-transmissable plasmid) or promote non-conjugational transmission (to become a phage like Mu).

It is certainly not novel to suggest that prokaryotic transposable elements behave in these ways, or to suggest that more frankly autonomous entities like phages have arisen from them[12-16,31]. However, we think it has not been sufficiently emphasized that non-phenotypic selection may inevitably give rise to transposable elements and that no phenotypic rationale for their origin and continued existence is thus required.

TRANSPOSABLE ELEMENTS IN EUKARYOTES

There has long been genetic evidence for the existence in eukaryotic genomes of transposable elements affecting phenotype[32]. These have been assigned roles in the regulation of eukaryotic gene expression and in evolution, but would have escaped genetic detection had they not had phenotypic effect. More recent evidence for transposable elements whose effects are not readily identified genetically has come fortuitously from studies of cloned eukaryotic DNAs. For instance, the Ty-1 element of yeast (which has no known phenotypic function) is flanked by direct repeats (like some prokaryotic transposons) and is transposable[33]. It is present in some 35 dispersed copies and comprises some 2% of the yeast genome (like a higher-eukaryotic middle-repetitive DNA). The directly repeated δ-sequence elements flanking it are found at still other sites (just as prokaryotic insertion sequences can be found flanking transposons or independently elsewhere in the genome). Cameron et al. suggest that 'Ty-1 may be a nonviral "parasitic" DNA' but then go on to suggest, we think unnecessarily, that transposition 'allows adaptation of a particular cell to a new environment' (ref. 33). The repetitive elements 412, copia and 297 of Drosophila are physically similar to Ty-1 (and to bacterial transposable elements) and are transposable[34-37]. Strobel et al. suggest 'it is possible that the sole function of these elements is to promote genetic variability, and that their gene products may only be necessary for the maintenance and mobility of the elements themselves, rather than for other cellular processes' (ref. 37). But if maintenance and mobility mechanisms exist, then no cellular function at all need be postulated.

A large fraction of many eukaryotic genomes consists of middle-repetitive DNAs, and the variety and patterns of their interspersion with unique sequence DNA make no particular phylogenetic or phenotypically functional sense. Britten, Davidson and collaborators have elaborated models which ascribe regulatory functions to middle-repetitive DNAs, and evolutionary advantage (in terms of adaptability) to the quantitative and qualitative changes in middle-repetitive DNA content observed even between closely related species[17,38-40]. Middle-repetitive DNAs are more conserved in sequence during evolution than are unique-sequence DNAs not coding for protein, and Klein *et al.* suggest that 'restraint on repetitive sequence divergence, either within the repeat families of a given species, or over evolutionary time spanning the emergence of different species, could be due to [phenotypic] selective pressures which prevent free sequence change in a large fraction of the repeat family members. Or perhaps repetitive sequences diverge as rapidly as do other sequences, but the type sequence of the family is preserved by frequent remultiplication of the "correct" surviving sequences' (ref. 41). The evidence for a phenotypically functional role for middle-repetitive sequences remains dishearteningly weak[40-43], and if the calculations of Kimura[44] and Salser and Isaacson[45] are correct, middle-repetitive DNAs together comprise too large a fraction of most eukaryotic genomes to be kept accurate by darwinian selection operating on organismal phenotype. The most plausible form of "remultiplication of the 'correct' surviving sequences" is transposition. If we assume middle-repetitive DNAs in general to be transposable elements or degenerate (and no longer transposable and ultimately to be eliminated) descendants of such elements, then the observed spectra of sequence divergence within families and changes in middle-repetitive DNA family sequence and abundance can all be explained as the result of non-phenotypic selection within genomes. No cellular function at all is required to explain either the behaviour or the persistence of middle-repetitive sequences as a class.

THE REST OF THE EUKARYOTIC GENOME

Middle-repetitive DNA can comprise more than 30% of the genome of a eukaryotic cell[46]. Another 1–40% consists of simple reiterated sequences whose functions remain unclear[10], and Smith has argued that 'a pattern of tandem repeats is the natural state of DNA whose sequence is not maintained by selection' (ref. 47). Even unique-sequence eukaryotic DNA consists in large part of elements which do not seem to be constrained by phenotypic selection pressures[45]. Some authors have argued that the intervening sequences which interrupt many eukaryotic structural genes are insertion sequence-like elements[6,48,49]. If they are, they are likely to be the degenerate and no-longer transposable descendants of transposable sequences whose insertion was rendered non-lethal by pre-existing cellular RNA : RNA splicing mechanisms. Such elements, once inserted, are relatively immune to deletion (since only very precise deletion can be non-lethal), and need retain only those sequence components required for RNA splicing. The rest of the element is free to drift and one expects (and observes) that only the position and number of intervening sequences in a family of homologous genes remain constant during evolution. Although evolutionary and regulatory phenotypic functions have been ascribed to intervening sequences[6,49-51], it is unnecessary to postulate any cellular function at all if these elements are indeed degenerate transposable elements arising initially from non-phenotypic selection. Another explanation for the origin and continued existence of intervening sequences, which also does not require phenotypically or evolutionarily advantageous roles, has been suggested elsewhere[50,51].

WHY DO PROKARYOTES AND EUKARYOTES DIFFER?

It is generally believed that prokaryotic genomes consist almost entirely of unique-sequence DNA maintained by phenotypic selection, whereas the possession of 'excess' unique and repetitive DNA sequences whose presence is at least difficult to rationalize in phenotypic terms is characteristic of eukaryotes. However, it is more accurate to say that there is a continuum of excess DNA contents; at least 1% of the *E. coli* genome can be made up of copies of six identified insertion sequences alone[16]. Yeast, whose genome is no larger than that of some prokaryotes, has few repeated sequences other than those coding for stable RNAs, and *Aspergillus* may have none[52,53]. There is in general (but with many exceptions) a positive correlation between excess DNA content, genome size and what we anthropocentrically perceive as 'evolutionary advancement'. Many interpret this as the cause and/or consequence of the increasing phenotypic complexity which characterizes organismal evolution, and attribute to excess DNA a positive role in the evolutionary process[17-19,40]. The interplay of phenotypic and non-phenotypic forces, and the importance of understanding both in attempts to restore the 'C-value paradox' are discussed more thoroughly by Orgel and Crick in the following article.[54]

There is another, simpler and perhaps obvious explanation. Non-phenotypic selection produces excess DNA, and excess DNA logically must be an energetic burden; phenotypic selection should favour its elimination[24,25]. The amount of excess (and hence total) DNA in an organism should be loosely determined by the relative intensities of the two opposing sorts of selection. The intensity of non-phenotypic pressure on DNA to survive even without function should be independent of organismal physiology. The intensity of phenotypic selection pressure to eliminate excess DNA is not, this being greatest in organisms for which DNA replication comprises the greatest fraction of total energy expenditure. Prokaryotes in general are smaller and replicate themselves and their DNA more often than eukaryotes (especially complex multicellular eukaryotes). Phenotypic selection pressure for small 'streamlined' prokaryotic genomes with little excess DNA may be very strong.

NECESSARY AND UNNECESSARY EXPLANATIONS

We do not deny that prokaryotic transposable elements or repetitive and unique-sequence DNAs not coding for protein in eukaryotes may have roles of immediate phenotypic benefit to the organism. Nor do we deny roles for these elements in the evolutionary process. We do question the almost automatic invocation of such roles for DNAs whose function is not obvious, when another and perhaps simpler explanation for their origin and maintenance is possible. It is inevitable that natural selection of the special sort we call non-phenotypic will favour the development within genomes of DNAs whose only 'function' is survival within genomes. When a given DNA, or class of DNAs, of unproven phenotypic function can be shown to have evolved a strategy (such as transposition) which ensures its genomic survival, then no other explanation for its existence is necessary. The search for other explanations may prove, if not intellectually sterile, ultimately futile.

We thank L. Bonen, R. M. MacKay and M. Schnare for help in development of the ideas presented here, and C. W. Helleiner, M. W. Gray, C. Stuttard, R. Singer, S. D. Wainwright and E. Butz for critical discussions.

We are especially grateful to C. E. Orgel and F. H. C. Crick for discussing with us the ideas presented in the following article before publication and for encouragement and support.

1. Dawkins, R. *The Selfish Gene* (Oxford University Press, 1976).
2. Crow, J. F. *Scient. Am.* **240**, 134–146 (1979).
3. Cox, E. C. & Gibson, T. C. *Genetics* **77**, 169–184 (1974).
4. Maynard Smith, J. *The Evolution of Sex* (Cambridge University Press, 1978).
5. Darnell, J. E. *Prog. Nucleic Acid Res. molec. Biol.* **22**, 327–353 (1978).
6. Crick, F. H. C. *Science* **204**, 264–271 (1979).
7. Murray, V. & Holliday, R. *FEBS Lett.* **106**, 5–7 (1979).
8. Sogin, M. L., Pace, B. & Pace, N. R. *J. biol. Chem.* **252**, 1350–1357 (1977).
9. Fedoroff, N. *Cell* **16**, 697–710 (1979).
10. John B. & Miklos, G. L. G. *Int. Rev. Cytol.* **58**, 1–114 (1979).
11. Zuckerkandl, E. *J. molec. Evol.* **9**, 73–122 (1976).
12. Cohen, S. N. *Nature* **263**, 731–738 (1976).
13. Starlinger, P. & Saedler, H. *Curr. Topics Microbiol. Immun.* **75**, 111–152 (1976).
14. Nevers, P. & Saedler, H. *Nature* **268**, 109–115 (1977).
15. Kleckner, N. *Cell* **11**, 11–23 (1977).
16. Kopecko, D. in *Plasmids and Transposons: Environmental Effects and Maintenance Mechanisms* (eds Stuttard, C. & Rozee, K. R.) (Academic New York, in the press).
17. Britten, R. J. & Davidson, E. H. *Q. Rev. Biol.* **46**, 111–138 (1971).
18. Ohno, S. *Evolution by Gene Duplication* (Springer, New York, 1970).
19. Hinegardner, R. in *Molecular Evolution* (ed. Ayala, F. J.) 160–199 (Sinauer, Sunderland, 1976).
20. Gould, S. J. & Lewontin, R. C. *Proc. R. Soc.* **B205**, 581–598 (1979).
21. Orgel, L. *Proc. R. Soc.* **B205**, 435–442 (1979).
22. Bodmer, W. in *Human Genetics: Possibilities and Realities*, 41–42 (Excerpta Medica, Amsterdam, 1979).
23. Johnsrod, L. *Molec gen. Genet.* **169**, 213–218 (1979).
24. Zamenhof, S. & Eichorn, H. H. *Nature* **216**, 456–458 (1967).
25. Koch, A. L. *Genetics* **72**, 297–316 (1972).
26. Shapiro, J. A. *Proc. natn. Acad. Sci. U.S.A.* **76**, 1933–1937 (1979).
27. Arthur, A. & Sherratt, D. *Molec. gen. Genet.* **175**, 267–274 (1979).
28. Gill, R., Heffron, F., Dougan, G. & Falkow, S. J. *Bact.* **136**, 742–756 (1978).
29. MacHattie, L. A. & Shapiro, J. A. *Proc. natn. Acad. Sci. U.S.A.* **76**, 1490–1494 (1979).
30. Huberman, P., Klaer, R., Kühn, S. & Starlinger, P. *Molec. gen. Genet.* **175**, 369–373 (1979).
31. Campbell, A. in *Biological Regulation and Development* Vol. 1 (ed. Goldberger, R. F.) 19–55 (Plenum, New York, 1979).
32. McClintock, B. *Cold Spring Harb. Symp. quant. Biol.* **16**, 13–47 (1952).
33. Cameron, J. R., Loh, E. Y & Davis, R. W. *Cell* **16**, 739–751 (1979).
34. Finnegan, D. J., Rubin, G. M., Young, M. W. & Hogness, D. S. *Cold Spring Harb. Symp. quant. Biol.* **42**, 1053–1063 (1977).
35. Carlson, M. & Brutlag, D. *Cell* **15**, 733–742 (1978).
36. Potter, S. S., Borein, W. J. Jr, Dunsmuir, P. & Rubin, G. M. *Cell* **17**, 415–427 (1979).
37. Strobel, E., Dunsmuir, P. & Rubin, G. M. *Cell* **17**, 429–439 (1979).
38. Britten, R. J. & Davidson, E. H. *Science* **165**, 349–357 (1969).
39. Davidson, E. H., Klein, W. H. & Britten, R. J. *Devl Biol.* **55**, 69–84 (1977).
40. Davidson, E. H. & Britten, R. J. *Science* **204**, 1052–1059 (1979).
41. Klein, W. H. *et al. Cell* **14**, 889–900 (1978).

42. Sheller, R. H., Constantini, F. D., Dozlowski, M. R., Britten, R. J. & Davidson, E. H. *Cell* **15**, 189–203 (1978).
43. Kuroiwa, A. & Natori, S. *Nucleic Acids Res.* **7**, 751–754 (1979).
44. Kimura, M. *Nature* **217**, 624–626 (1968).
45. Salser, W. & Isaacson, J. S. *Prog. Nucleic Acids Res. molec. Biol.* **19**, 205–220 (1976).
46. Lewin, B. *Cell* **4**, 77–93 (1975).
47. Smith, G. P. *Science* **191**, 528–534 (1976).
48. Tsujimoto, Y. G. & Suzuki, Y. *Cell* **18**, 591–600 (1979).
49. Gilbert, W. *Nature* **271**, 501 (1978).
50. Doolittle, W. F. *Nature* **272**, 581–582 (1978).
51. Darnell, J. E. *Science* **202**, 1257–1260 (1978).
52. Roberts, T. M., Lauer, G. D. & Klotz, L. *CRC Crit. Rev. Biochem.* **3**, 349–451 (1976).
53. Timberlake, W. F. *Science* **702**, 973–974 (1978).
54. Orgel, G. E. & Crick, F. H. C. *Nature* **284**, 604–607 (1980).

W. Ford Doolittle and Carmen Sapienza are in the Department of Biochemistry, Dalhousie University, Halifax, Nova Scotia.

This article first appeared in *Nature* Vol. **284**, pp. 601–603; 1980.

Selfish DNA: the ultimate parasite

L. E. ORGEL AND F. H. C. CRICK

The DNA of higher organisms usually falls into two classes, one specific and the other comparatively nonspecific. It seems plausible that most of the latter originated by the spreading of sequences which had little or no effect on the phenotype. We examine this idea from the point of view of the natural selection of preferred replicators within the genome.

THE object of this short review is to make widely known the idea of selfish DNA. A piece of selfish DNA, in its purest form, has two distinct properties:

(1) It arises when a DNA sequence spreads by forming additional copies of itself within the genome.
(2) It makes no specific contribution to the phenotype.

This idea is not new. We have not attempted to trace it back to its roots. It is sketched briefly but clearly by Dawkins[1] in his book *The Selfish Gene* (page 47). The extended discussion (pages 39–45) after P. M. B. Walker's article[2] in the CIBA volume based on a Symposium on Human Genetics held in June 1978 shows that it was at that time already familiar to Bodmer, Fincham and one of us. That discussion referred specifically to repetitive DNA because that was the topic of Walker's article, but we shall use the term selfish DNA in a wider sense, so that it can refer not only to obviously repetitive DNA but also to certain other DNA sequences which appear to have little or no function, such as much of the DNA in the introns of genes and parts of the DNA sequences between genes. The catch-phrase 'selfish DNA' has already been mentioned briefly on two occasions[3,4]. Doolittle and Sapienza[5] (see the previous article) have independently arrived at similar ideas.

THE AMOUNT OF DNA

The large amounts of DNA in the cells of most higher organisms and, in particular, the exceptionally large amounts in certain animal and plant species—the so-called C value paradox—has been an unsolved puzzle for a considerable period (see reviews in refs 6–8). As is well known, this

DNA consists in part of 'simple' sequences, an extreme example of which is the very large amounts of fairly pure poly d(AT) in certain crabs. Simple sequences, which are situated in chromosomes largely but not entirely in the heterochromatin, are usually not transcribed. Another class of repetitive sequences, the so-called 'intermediate repetitive', have much longer and less regular repeats. Such sequences are interspersed with 'unique' DNA at many places in the chromosome, the precise pattern of interspersion being to some extent different in different species. Leaving aside genes which code for structural RNA of one sort or another (such as transfer RNA and ribosomal RNA), which would be expected to occur in multiple copies (since, unlike protein, their final products are the result of only one stage of magnification, not two), the majority of genes coding for proteins appear to exist in 'single' copies, meaning here one or a few. A typical example would be the genes for α-globin, which occur in one to three copies and the human β-like globins, of which there are four main types, all related to each other but used for slightly different purposes. Notable exceptions are the proteins of the immune system, and probably those of the histocompatibility and related systems. Another exception is the genes for the five major types of histone which also occur in multiple copies. Even allowing for all such special case, the estimated number of genes in the human genome appears too few to account for the 3×10^9 base pairs found per haploid set of DNA, although it must be admitted that all such arguments are very far from conclusive.

Several authors [8-13] have suggested that the DNA of higher organisms consists of a minority of sequences with highly specific functions plus a majority with little or no specificity. Even some of the so-called single-copy DNA may have no specific function. A striking example comes from the study of two rather similar species of *Xenopus*. These can form viable hybrids, although these hybrids are usually sterile. However, detailed molecular hybridization studies show that there has been a large amount of DNA sequence divergence since the evolutionary separation of their forebears. These authors [13] conclude 'only one interpretation seems reasonable, and that is that the specific sequence of much of the single-copy DNA is not functionally required during the life of the animal. This is not to say that this DNA is functionless, only that its specific sequence is not important'.

There is also evidence to suggest that the majority of DNA sequences in most higher organisms do not code for protein since they do not occur at all in messenger RNA (for reviews see refs 14, 15). Nor is it very plausible that all this extra DNA is needed for gene control, although some portion of it certainly must be.

We also have to account for the vast amount of DNA found in certain species, such as lilies and salamanders, which may amount to as much as 20 times that found in the human genome. It seems totally implausible that the number of radically different genes needed in a salamander is 20 times that in a man. Nor is there evidence to support the idea that salamander genes are mostly present in about 20 fairly similar copies. The conviction has been growing that much of this extra DNA is 'junk', in other words, that it has little specificity and conveys little or no selective

between sps. is it the junk that differs
or specific sequences.

advantage to the organism.

Another place where there appears to be more nucleic acid than one might expect is in the primary transcripts of the DNA of higher organisms which are found in the so-called heteronuclear RNA. It has been known for some time that this RNA is typically longer than the messenger RNA molecules found in the corresponding cytoplasm. Heteronuclear RNA contains these messenger RNA sequences but has many other sequences which are never found in the cytoplasm. The phenomenon has been somewhat clarified by the recent discovery of introns in many genes (for a general introduction see ref. 4). Although the evidence is still very preliminary, it certainly suggests that much of the base sequence in the interior of some introns may be junk, in that these sequences drift rapidly in evolution, both in detail and in size. Moreover, the number of introns may differ even in closely related genes, as in the two genes for rat preproinsulin[16]. Whether there is junk between genes is unclear but it is noteworthy that the four genes for the human β-like globins, which occur fairly near together in a single stretch of DNA, occupy a region no less than 40 kilobases long[17]. This greatly exceeds the total length of the four primary transcripts (that is the four mRNA precursors), an amount estimated to be considerably less than 10 kilobases. There is little evidence to indicate that there are other coding sequences between these genes (although the question is still quite open) and a tenable hypothesis is that much of this interspersed DNA has little specific function.

In summary, then, there is a large amount of evidence which suggests, but does not prove, that much DNA in higher organisms is little better than junk. We shall assume, for the rest of this article, that this hypothesis is true. We therefore need to explain how such DNA arose in the first place and why it is not speedily eliminated, since, by definition, it contributes little or nothing to the fitness of the organism.

WHAT IS SELFISH DNA?

The theory of natural selection, in its more general formulation, deals with the competition between replicating entities. It shows that, in such a competition, the more efficient replicators increase in number at the expense of their less efficient competitors. After a sufficient time, only the most efficient replicators survive. The idea of selfish DNA is firmly based on this general theory of natural selection, but it deals with selection in an unfamiliar context.

The familiar neo-darwinian theory of natural selection is concerned with the competition between organisms in a population. At the level of molecular genetics it provides an explanation of the spread of 'useful' genes or DNA sequences within a population. Organisms that carry a gene that contributes positively to fitness tend to increase their representation at the expense of organisms lacking that gene. In time, only those organisms that carry the useful gene survive. Natural selection also predicts the spread of a gene or other DNA sequence within a single genome, provided certain conditions are satisfied. If an organism carrying several copies of the sequence is fitter than an organism carrying a single copy, and if mechanisms exist for the multiplication of the relevant sequence, then natural selection must lead to the emergence of a population in which the sequence is represented several times in every genome.

The idea of selfish DNA is different. It is again concerned with the spread of a given DNA within the genome. However, in the case of selfish DNA, the sequence which spreads makes no contribution to the phenotype of the organism, except insofar as it is a slight burden to the cell that contains it. Selfish DNA sequences may be transcribed in some cases and not in others. The spread of selfish DNA sequences within the genome can be compared to the spread of a not-too-harmful parasite within its host.

MECHANISMS FOR DNA SPREADING

The inheritance of a repeated DNA sequence in a population of eukaryotes clearly requires that the multiplication which produced it occurred in the germ line. Furthermore, any mechanism that can lead to the multiplication of useful DNA will probably lead to the multiplication of selfish DNA (and vice versa). Of course, natural selection subsequently discriminates between multiple sequences of different kinds, but it does not necessarily prevent the multiplication of neutral or harmful sequences.

Multiplication in the germ-line sequence can occur in non-dividing cells or during meiosis and mitosis (within lineages that lead to the germ line). In the former case, the mechanisms available resemble those that are well documented for prokaryotes, that is, multiplication may occur in eukaryotes through the integration of viruses or of elements analogous to transposons and insertion sequences. Doolittle and Sapienza[5] have discussed these mechanisms in some detail, particularly for prokaryotes. They are likely to lead to the spreading of DNA sequences to widely separated positions on the chromosomes.

During mitosis and meiosis, multiplication (or deletion) is likely to occur by unequal crossing over. This mechanism will often lead to the formation of tandem repeats. It is well documented for the tRNA 'genes' of *Drosophila* and for various other tandemly repeated sequences in higher organisms.

THE AMOUNT AND LOCATION OF SELFISH DNA

Natural selection 'within' the genome will favour the indefinite spreading of selfish preferred replicators. Natural selection between genotypes provides a balancing force that attempts to maintain the total amount of selfish DNA at an equilibrium (steady state) level—organisms whose genomes contain an excessive proportion of selfish DNA would be at a metabolic disadvantage relative to organisms with less selfish DNA, and so would be eliminated by the normal mechanism of natural selection. Excessive spreading of functionless replicators may be considered as a 'cancer' of the genome—the uncontrolled expansion of one segment of the genome would ultimately lead to the extinction of the genotype that permits such expansion. Of course, we do not know whether extinction of genotypes in nature ever occurs for this reason.

It is hard to get beyond generalities of this kind. To do so we would, at least, need to know how much selective disadvantage results from the presence of a given amount of useless DNA. Even this minimal information is not easily acquired, so we cannot produce other than qualitative arguments.

It seems certain that the metabolic energy cost of replicating a superfluous short DNA sequence in a genome containing 10^9 base pairs would be very small. If, for example, the selective disadvantage were equal to the proportion of the genome made up by the extra DNA, a sequence of 1,000 base pairs would produce a selective disadvantage of only 10^{-6}. If the selective disadvantage were proportional to the extra energy cost divided by the total metabolic energy expended per cell per generation, the disadvantage would be much smaller. The selective disadvantage might be greater in more stringent conditions, but it is still hard to believe that a relatively small proportion of selfish DNA could be selected against strongly.

On the other hand, when the total amount of selfish DNA becomes comparable to or greater than that of useful DNA, it seems likely that the selective disadvantage would be significant. We may expect, therefore, that the mechanisms for the formation and deletion of nonspecific DNA will adjust, in each organism, so that the load of DNA is sufficiently small that it can be accommodated without producing a large selective disadvantage. The proportion of nonspecific DNA in any particular organism will thus depend on the lifestyle of the organism, and particularly on its sensitivity to metabolic stress during the most vulnerable part of the life cycle.

We can make one prediction on the basis of energy costs. Selfish DNA will accumulate to a greater extent in non-transcribed regions of the genome than in those that are transcribed. Of course, selfish DNA will in most cases be excluded from translated sequences, because the insertion of amino acids within a protein will almost always have serious consequences, even in diploid organisms (but see the suggestion by F.H.C.C.[18]).

At first sight it might seem anomalous that natural selection does not eliminate all selfish DNA. Since the suggestion that much eukaryotic DNA is useless distinguishes the selfish DNA hypothesis from many closely related proposals, it may be useful to take up this point in some detail.

First, the elimination of disadvantaged organisms from a population, by their more favoured competitors, takes a number of generations several times larger than the reciprocal of the selective disadvantage. If the selective disadvantage associated with a stretch of useless DNA in higher organisms is only 10^{-6}, it would take 10^6–10^8 years to eliminate it by competition. For typical higher organisms this is a very long time, so the elimination of a particular stretch of selfish DNA may be a very slow process even on a geological time scale. Second, the mechanisms for the deletion of short sequences of DNA may be inefficient, since there is no strong selective pressure for the development of 'corrective' measures when the 'fault' carries a relatively small selective penalty. Taken together, these arguments suggest that the elimination of a particular piece of junk from the genome may be a very slow process.

This in turn suggests that the amount of useless DNA in the genome is a consequence of a dynamic balance. The organism 'attempts' to limit the spread of selfish DNA by controlling the mechanism for gene duplication, but is constrained by imperfections in genetic processes and/or by the need to permit some duplication of advantageous genes. Selfish DNA sequences 'attempt' to subvert these mechanisms and may be able to do so comparatively rapidly because mutation will affect them directly. On the other hand, the defence mechanisms of the host are likely to depend on the action of protein and therefore may evolve more slowly. Once established within the genome, useless sequences, probably have a long 'life expectancy'.

For any particular type of selfish DNA, there is no reason that a steady state should necessarily be reached in evolution. The situation would be continually changing. A particular type of DNA might first spread rather successfully over the chromosomes. The host might then evolve a mechanism which reduced or eliminated further spreading. It might also evolve a method for preferentially deleting it. At the same time, random mutations in the selfish DNA might make it more like ordinary DNA and so, perhaps, less easy to remove. Eventually, these sequences, possibly by now rather remote from those originally introduced, may cease to spread and be slowly eliminated. Meanwhile, other types of selfish DNA may originate, expand and evolve in a similar way.

In short, we may expect a kind of molecular struggle for existence within the DNA of the chromosomes, using the process of natural selection. There is no reason to believe that this is likely to be much simpler or more easy to predict than evolution at any other level. At bottom, the existence of selfish DNA is possible because DNA is a molecule which is replicated very easily and because selfish DNA occurs in an environment in which DNA replication is a necessity. It thus has the opportunity of subverting these essential mechanisms to its own purpose.

THE INHERITANCE OF SELFISH DNA

Although the inheritance of selfish DNA will occur mainly within a mendelian framework, it is likely to be different in detail and more complex than simple mendelian inheritance. This is due both to the multiplication mechanisms, which in one way or another will produce repeated copies (see the discussion by Doolittle and Sapienza[5]), and to the fact that these copies are likely to be distributed round the chromosomes rather than being located in a single place in the genome as most normal genes are. For both these reasons, a particular type of selfish DNA is likely to spread more rapidly through a population than would a normal gene with a low selective advantage. It will be even more rapid if selfish DNA can spread horizontally between different individuals in a population, due to viruses or other infectious agents, although it should be remembered that such 'infection' must affect the germ line and not merely the soma. If this initial spread takes place when the additional DNA produced is relatively small in amount, it is unlikely to be seriously hindered by the organism selecting against it. The study of these processes will clearly require a new type of population genetics.

CAN SELFISH DNA ACQUIRE A SPECIFIC FUNCTION?

It would be surprising if the host organism did not occasionally find some use for particular selfish DNA sequences, especially if there were many different sequences widely distributed over the chromosomes. One obvious use, as repeatedly stressed by Britten and Davidson[19,20], would be for control purposes at one level or another. This seems more than plausible.

It has often been argued (see, for example, ref. 21) that for the evolution of complex higher organisms, what is required is not so much the evolution of new proteins as the evolution of new control mechanisms and especially mechanisms which control together sets of genes which previously had been regulated separately. To be useful, a new control sequence of the DNA is likely to be needed in a number of distinct places in the genome. It has rarely been considered how this could be brought about expeditiously by the rather random methods available to natural selection.

A mechanism which scattered, more or less at random, many kinds of repeated sequences in many places in the genome would appear to be rather good for this purpose. Most sets of such sequences would be unlikely to find themselves in the right combination of places to be useful but, by chance, the members of one particular set might be located so that they could be used to turn on (or turn off) together a set of genes which had never been controlled before in a coordinated way. A next way of doing this would be to use as control sequences not the many identical copies distributed over the genome, but a small subset of these which had mutated away from the master sequence in the same manner.

On this picture, each set of repeated sequences might be 'tested' from time to time in evolution by the production of a control macromolecule (for example, a special protein) to recognize those sequences. If this produced a favourable result, natural selection would confirm and extend the new mechanism. If not, it would be selected against and discarded. Such a process implies that most sets of repeated sequences will never be of use since, on statistical grounds, their members will usually be in unsuitable places.

It thus seems unlikely that all selfish DNA has acquired a special function, especially in those organisms with very high C values. Nor do we feel that if one example of a particular sequence acquires a function, all the copies of that sequence will necessarily do so. As selfish DNA is likely to be distributed over the chromosomes in rather a random manner, it seems unlikely that every copy of a potentially useful sequence will be in the right position to function correctly. For example, if a specific sequence within an intron were used to control the act of splicing that intron, a similar sequence in an untranscribed region between genes would obviously not be able to act in this way.

In some circumstances, the sheer bulk of selfish DNA may be used by the organism for its own purpose. That is, the selfish DNA might acquire a nonspecific function which gives the organism a selective advantage. This is the point of view favoured by Cavalier-Smith in a very detailed and suggestive article[12] which the reader should consult. He proposes that excess DNA may be the mechanism the cell uses to slow up development or to make bigger cells. However, we suspect that both slow growth and large cell size could be evolved just as well by other more direct mechanisms. We prefer to think that the organism has tolerated selfish DNA which has arisen because of the latter's own selective pressure.

Thus, some selfish DNA may acquire a useful function and confer a selective advantage on the organism. Using the analogy of parasitism, slightly harmful infestation may ultimately be transformed into a symbiosis. What we would stress is that not all selfish DNA is likely to become useful. Much of it may have no specific function at all. It would be folly in such cases to hunt obsessively for one. To continue our analogy, it is difficult to accept the idea that all human parasites have been selected by human beings for their own advantage.

LIFE STYLE

The effect of nonspecific DNA on the life style of the organism has been considered by several authors, in particular by Cavalier-Smith[12] and by Hindergardner[8]. We shall not attempt to review all their ideas here but instead will give one example to show the type of argument used.

Bennett[22] has brought together the measurements of DNA content for higher herbaceous plants. There is a striking connection between DNA content per cell and the minimum generation time of the plant. In brief, if such an angiosperm has more than 10 pg of DNA per cell, it is unlikely to be an ephemeral (that is, a plant with a short generation time). If it is a diploid and has more than 30 pg of DNA, it is highly likely to be an obligate perennial, rather than an annual or an ephemeral. The converse, however, is not true, there being a fair number of perennials with a DNA content of less than 30 pg and a few with less than 10 pg. A clear picture emerges that if a herbaceous plant has too much DNA it cannot have a short generation time.

This is explained by assuming that the extra DNA needs a bigger nucleus to hold it and that this increases both the size of the cell and the duration of meiosis and generally slows up the development of the plant. An interesting exception is that the duration of meiosis, is, if anything, shorter for polyploid species than for their diploid ancestors[23]. This suggests that it is the ratio of good DNA to junk DNA rather than the total DNA content which influences the duration of meiosis.

An analogous situation may obtain in certain American species of salamander. These often differ considerably in the rapidity of their development and of their life cycles, the tropical species tending to take longer than the more temperate ones. Drs David Wake and Herbert MacGregor (personal communciation) tell us that preliminary evidence suggests that species with the longer developmental times often have the higher C values. This appears to parallel the situation just described for the herbaceous plants. It remains to be seen if further evidence will continue to support this generalization. (See the interesting paper by Oeldorfe et al.[25] on 25 species of frogs. They conclude that 'genome size sets a limit beyond which development cannot be accelerated'.)

TESTING THE THEORY

The theory of selfish DNA is not so vague that it cannot be tested. We can think of three general ways to do this. In the first place, it is important to know where DNA sequences occur which appear to have little obvious function, whether they are associated with flanking or other sequences of any special sort and how homologous sequences differ in different organisms and in different species, either in sequence or in position on the chromosome. For example, it has recently been shown by Young[24] that certain intermediate repetitive sequences in *Drosophila* are often in different chromosomal positions in different strains of the same species.

Second, if the increase of selfish DNA and its movement around the chromosome are not rare events in evolution, it may be feasible to study, in laboratory experiments, the actual molecular mechanisms involved in these processes.

Third, one would hope that a careful study of all the nonspecific effects of extra DNA would give us a better idea of how it affected different aspects of cellular behaviour. In particular, it is important to discover whether the addition of nonspecific DNA does, in fact, slow down cells metabolically and for what reasons. Such information, together with a careful study of the physiology and life style of related organisms with dissimilar amounts of DNA, should eventually make it possible to explain these differences in a convincing way.

CONCLUSION

Although it is an old idea that much DNA in higher organisms has no specific function[8-12], and although it has been suggested before that this nonspecific DNA may rise to levels which are acceptable or even advantageous to an organism[8,12], depending on certain features of its life style, we feel that to regard much of this nonspecific DNA as selfish DNA is genuinely different from most earlier proposals. Such a point of view is especially useful in thinking about the dynamic aspects of nonspecific DNA. It directs attention to the mechanisms involved in the spread and evolution of such DNA and it cautions one against looking for a special function for every piece of DNA which drifts rapidly in sequence or in position on the genome.

While proper care should be exercised both in labelling as selfish DNA every piece of DNA whose function is not immediately apparent and in invoking plausible but unproven hypotheses concerning the details of natural selection, the idea seems a useful one to bear in mind when exploring the complexities of the genomes of higher organisms. It could well make sense of many of the puzzles and paradoxes which have arisen over the last 10 or 15 years. The main facts are, at first sight, so odd that only a somewhat unconventional idea is likely to explain them.

We thank W. Ford Doolittle and C. Sapienza for showing us their article before publication, and Drs D. Wake and H. MacGregor for allowing us to quote some of their unpublished conclusions about salamanders. This work was supported by the Eugene and Estelle Ferhauf Foundation, the J. W. Kieckhefer Foundation, the Ahmanson Foundation and the Samuel Roberts Noble Foundation.

1. Dawkins, R. *The Selfish Gene* (Oxford University Press, 1976).
2. Walker, P. M. B. in *Human Genetics: Possibilities and Realities,* 25–38 (Excerpta Medica, Amsterdam, 1979).
3. Crick, F. H. C. in *From Gene to Protein: Information Transfer in Normal and Abnormal Cells* (eds Russell, T. R., Brew, K., Faber, H. & Schultz, J.) 1–13 (Academic, New York, 1979).
4. Crick, F. H. C. *Science* **204**, 264–271 (1979).
5. Doolittle, W. F. & Sapienza, C. *Nature* **284**, 601–603 (1980).
6. Callan, H. G. *J. Cell Sci.* **2**, 1–7 (1967).
7. Thomas, C. A. *A. Rev. Genet.* **5**, 237–256 (1971).
8. Hinegardner, R. in *Molecular Evolution* (ed. Ayala, F. J.) 179–199 (Sinauer, Sunderland, 1976).
9. Commoner, B. *Nature* **202**, 960–968 (1964).
10. Ohno, S. *J. hum. Evolut.* **1**, 651–662 (1972).
11. Comings, D. E. *Adv. hum. Genet.* **3**, 237–436 (1972).
12. Cavalier-Smith, T. *J. Cell Sci.* **34**, 274–278 (1978).
13. Galan, G. A., Chamberlain, M. E., Hough, B. R., Britten, R. J. & Davidson, E. H. in *Molecular Evolution* (ed. Ayala, F. J.) 200–224 (Sinauer, Sunderland, 1976).
14. Bishop, J. O. *Cell* **2**, 81–86 (1974).
15. Lewin, B. *Cell* **4**, 11–20 (1975); *Cell* **4**, 77–93 (1975).
16. Lomedico, P. *et al. Cell* **18**, 545–558 (1979).
17. Bernards, R., Little, P. F. R., Annison, G., Williamson, R. & Flavell, R. A. *Proc. natn. Acad. Sci. U.S.A.* **76**, 4827–4831 (1979).
18. Crick, F.H.C. *Eur. J. Biochem.* **83**, 1–3 (1978).
19. Britten, R. J. & Davidson, E. H. *Science* **165**, 349–358 (1969).
20. Davidson, E. U. & Britten, R. J. *Science* **204**, 1052–1059 (1979).
21. Wilson, A. C. in *Molecular Evolution* (ed. Ayala, F. J.) 225–236 (Sinauer, Sunderland, 1976).
22. Bennett, M. D. *Proc. R. Soc.* **B181**, 109–135 (1972).
23. Bennett, M. D. & Smith, J. B. *Proc. R. Soc.* **B181**, 81–107 (1972).
24. Young, M. W. *Proc. natn. Acad. Sci. U.S.A.* **76**, 6274–6278 (1979).
25. Oeldorfe, E., Nishioka, M. & Bachmann, K. *Sonderdr. Z.F. Zool, System Evolut.* **16**, 216–24 (1978).

L. E. Orgel and F. H. C. Crick are in the Salk Institute, La Jolla, California.

This article first appeared in *Nature* Vol. **284**, pp. 604–607; 1980.

Evolution of globin genes

A. J. JEFFREYS

GLOBIN genes are an ideal system for studying the molecular evolution of genes and multigene families. Globins are widespread in nature and include the tetrameric haemoglobins of higher vertebrates, monomeric haemoglobins of protochordates, a variety of invertebrate globins characterised in molluscs, the midge larva and the bloodworm, mono-meric myoglobins, and the monomeric leghaemoglobins found in the root nodules of nitrogen-fixing plants[1]. The amino acid sequences of many of these globins have been determined, and sequence homologies point to a common evolutionary origin of most or all globins. Detailed sequence comparisons of different species' globins have enabled phylo-genetic trees of globins to be constructed[1,2], and have shown that globin sequences tend to diverge in evolution at a constant rate, independent of the lineage being studied. This evolutionary molecular clock is typical of many protein sequences, and has proved an invaluable aid in determin-ing the time of species divergence or gene duplication (see Wilson *et al.*[3]).

Higher vertebrates code for a variety of globins; for example, man has eight active globin genes specifying haemoglobin polypeptides, which can be divided by sequence homology into two families. The α-globin related family consists of a ζ-globin gene expressed during early embryo-genesis and two almost-identical α-globin genes expressed in the foetus and adult. The β-related family contains a single embryonic ϵ-globin gene, two very similar foetal globin genes ($^{G}\gamma$ and $^{A}\gamma$), a minor adult δ-globin gene and the major adult β-globin gene. In addition, there are an unknown number of myoglobin genes. All of these gene products show significant amino acid sequence homology and, by using the molecular clock rate of globin sequence divergence, it is possible to deduce the timing of the gene duplications that gave rise to these families[2,4]. The most ancient duplication, about 1100 million years ago, gave rise to the ancestors of haemoglobin and myoglobin genes. The $\alpha\beta$-globin gene duplication occurred about 500 million years ago, early in the evolution of vertebrates. The β-globin gene family evolved more recently, with a foetal-adult duplication about 200 million years ago, an ϵ–γ duplication 100 million years ago and a δ–β duplication 40 million years ago.

Over the last four years, vertebrate haemoglobin genes have been investigated intensively using recombinant DNA techniques. This analysis has been facilitated by the isolation and purification *via* complementary DNA (cDNA) cloning of α- and β-globin messenger RNA sequences. Globin cDNA clones have been used to detect corresponding globin genes in Southern blot analyses of total genomic DNA cleaved with restriction endonucleases, and to isolate these genes from recombinant λ bacteriophage libraries. To date, α- and β-related globin genes have been studied in detail in animals ranging from amphibians to man and, so far, complete DNA sequences of some 15 different genes have been reported.

These studies at the DNA level have further shown the suitability of globin genes for evolutionary studies. Globin genes, in common with many other genes in higher eukaryotes, contain intervening sequences, the evolutionary history of which is beginning to be traced by comparative molecular studies. The α- and β-globin gene families have been found to be arranged in gene clusters that have probably arisen by tandem gene duplication. Evidence for genetic interchange between members of a gene cluster is accumulating. Additional inactive pseudo-genes, unsuspected from protein studies, have been found.

This paper reviews some of the recent advances in our understanding of gene evolution that have resulted from comparative molecular studies of globin genes.

EVOLUTION OF GLOBIN CODING SEQUENCES

By comparing the DNA sequences of the coding regions (exons) of globin genes, we can learn something of the rates and modes of exon evolution. Many such comparisons, both of duplicated, diverged genes within a single species, and of homologous genes in different species, have shown that nucleotide substitutions are not scattered at random throughout exons, but show a marked clustering at third codon positions, where they cause synonymous codon changes[4-6]. Clearly, selection has eliminated the bulk of substitutions that caused amino acid replacements (replacement site substitutions), whereas many synonymous or silent site substitutions probably have little or no effect on gene function and can be fixed in evolution. A similar phenomenon has been found in many other gene systems (see Jukes and King[7], Jukes[8]).

There have been several attempts to quantify the rates of replacement site and silent site substitution[4,6,9-11]. Replacement sites diverge in a clock-like fashion at about 0.1% per million years; this monotonous change of sequence is of course expected, in view of the protein evolutionary clock hypothesis plus the fact that most divergence times used by Perler et al.[6] and Efstratiadis et al.[4] were estimated from protein sequence data, including globin sequences. Replacement site divergences can be used interchangeably with amino acid sequence divergences to construct molecular phylogenies[4]. The clock rates for these divergences vary from protein to protein, and presumably reflect the proportion of amino acid sites that can be altered without causing loss of protein function. Invariant sites are likely to be important for protein function and in the globins include, in particular, residues involved in haem binding.

Silent sites in globin exons tend to diverge much more rapidly, at an initial rate of about 1% per million years (cf. 0.1%/million years for replacement sites[4,6,9]). It is not known whether this rate is constant in evolution and independent of lineage, since few closely related genes have been sequenced. There is tentative evidence from other gene comparisons to suggest that silent sites, unlike replacement sites, might evolve at a constant rate in all genes[6,9,10]. If so, then this silent site clock would be of great use for constructing phylogenies of closely related species.

If a silent site clock running at the same rate for all genes does exist, then this would support the idea, repeatedly proposed, that those silent site changes that are fixed in evolution are selectively neutral[7,8,10]. However, not all possible silent site replacements are necessarily neutral. In a comparison of human and rabbit β-globin mRNA sequences, Kafatos et al.[5] noted that silent changes were not scattered at random along the mRNA but were clustered into regions that tended also to be rich in replacement site substitutions. Similarly, Miyata and Yasunaga[11] find that silent site changes seem to accumulate less rapidly than nucleotide changes in possibly functionless pseudogenes (see below). Thus, some silent changes are eliminated in evolution, perhaps as a result of interfering with transcription or mRNA/precursor mRNA structure, processing or export. Alternatively, a silent substitution might generate a synonymous codon for which no abundant transfer RNA exists.

Perler et al.[6] and Efstratiadis et al.[4] find that the rate of silent site substitution apparently slows down after about 100 million years of divergence and thereafter proceeds at a rate similar to replacement site substitutions. It is still not certain whether this shift in rate is real or an artifact caused by analysing highly divergent sequences, or by the methods used for correcting these divergences for multiple substitutions (see Jeffreys[12], Kimura[10]). If the shift is real, then it suggests that there are (at least) two classes of silent site substitution. One occurs exceedingly rapidly, at about 1.5% per million years for globin, and might represent neutral sites that become fully randomised within 100 million years of divergence. The second set changes slowly (about 0.1%/million years) and might represent the gradual appearance of adaptive changes in gene expression, or might be driven by a slow shift in the codon utilisation pattern of the genome[13].

Perler et al.[6] have proposed that the replacement site substitution clock, and therefore amino acid divergence, is driven primarily by selection. They argue, that, since replacement sites diverge at about 10% of the rate of silent changes, 90% of replacement site substitutions are eliminated by selection. Thus, only 10% of possible replacement changes could be neutral, which suggests that replacement site divergence should saturate at 10%; this does not occur. This argument has not been accepted by Kimura[10].

EVOLUTION OF INTERVENING SEQUENCES IN GLOBIN GENES

A wide variety of α- and β-related globin genes have been analysed in detail in species including man, rabbit, mouse, chicken and *Xenopus laevis* (see Jeffreys[12]). All active vertebrate globin genes studied contain two intervening sequences interrupting the protein coding sequence. In every case, the intervening sequences occur at precisely homologous positions in the genes. As noted by Leder *et al.*[14], this establishes that the discontinuous structure must have been in existence at least 500 million years ago, before the $\alpha\beta$-globin gene duplication arose, and that no intervening sequence has subsequently been gained or lost by active vertebrate globin genes. This structure is also preserved in the $\psi\alpha 1$ pseudogene in man[15], in the rabbit $\psi\beta 2$ sequence[16] and in the goat β^x pseudogene[17].

A remarkable exception to this rule occurs in the mouse $\psi\alpha 3$ pseudogene, in which both intervening sequences have been precisely removed recently in evolution[18,19]. In addition, this pseudogene has accumulated a number of deletions, insertions and frameshifts sufficient to render the gene non-functional. It is not known whether the removal of the intervening sequences was initially responsible for silencing the gene. Precise intron loss has also been noted in the rat preproinsulin I gene, a functional gene that lacks one of the two introns seen in the closely related rat preproinsulin II gene as well as in human and chicken preproinsulin genes[6,20,21].

Jensen *et al.*[22] have recently analysed the structure of a leghaemoglobin (Lb) gene coded by soybean DNA. This gene has three intervening sequences, not two, but the first and third introns appear to be at positions homologous to the two vertebrate introns. The central intron in the Lb gene interrupts what is a continuous coding sequence in animal genes. The extraordinary thing is that there is any similarity at all between plant and animal globin genes. It has often been assumed that animal globins and Lb are the products of convergent evolution. However, the similar gene structure, as well as (limited) sequence homology, points to a common evolutionary origin of these globins. Since the vast majority of plants do not produce Lb, it is difficult to see how this relationship can be traced back to the common ancestor of plants and animals. Instead, it seems possible that horizontal gene transfer might have occurred between animals and plants, although sequence comparisons between vertebrate and invertebrate haemoglobins, myoglobins and Lb give no clue as to the source of the Lb gene[1]. It will be very interesting to see whether invertebrate globin genes, or vertebrate myoglobin genes, also contain three introns. If so, then it would seem likely that the central intron was eliminated in the lineage leading to the vertebrates at some time before the $\alpha\beta$-globin gene duplication.

How are introns removed so precisely in germ cell DNA during evolution? A variety of mechanisms has been postulated, including: recombination of a cDNA copy of globin mRNA into a split gene; annealing of mRNA to the coding strand during DNA replication, followed by excision of the displaced single-stranded intron DNA loops; direct action of the RNA splicing system on the anti-coding DNA strand at the replication fork; inclusion of a gene within a proretrovirus, followed by the production of spliced retroviral RNA and subsequent retroviral reintegration to produce an intron-less gene[18,19,23].

Comparisons of intron sequences in homologous globin genes have shown that, in contrast to coding sequences, intervening sequences evolve rapidly by base substitution and by small deletions and insertions[24-26]. These microdeletions/insertions might have been generated by slipped mispairing during DNA replication[4], and would not generally be tolerated in exons. No accurate estimate of the rate of intron divergence has yet been reported, nor have any conserved intron sequences been noted, except for the consensus sequences at splice junctions[27].

Despite the high frequency of microdeletions/insertions in globin genes during evolution, the lengths of introns have remained surprisingly constant; for example, the length of the first intervening sequence in mammalian α- and β-globin genes lies within the range of 116 to 130 base-pairs, despite the extreme age of the $\alpha\beta$-globin gene duplication (see Jeffreys[12]). Van Den Berg et al.[24] have suggested that intron length, rather than sequence, might be important in some way for globin gene function. However, rapid evolutionary changes in intron lengths have occurred in other genes, such as preproinsulin genes[6], vitellogenin genes[28], δ-crystallin genes[29] and ovalbumin-related genes[30].

The exons in vertebrate globin genes appear to correlate with domains in haemoglobin, as first proposed in general for split genes by Blake[31]. The central exon codes for the haem-binding domain of globin, although the other two exon products are required to maintain a stable haem-protein complex[32,33]. The distribution of inter-subunit contacts in haemoglobin appears to be non-random with respect to the three exons[34]. At first glance, the finding of an extra intron in the haem-binding exon of the leghaemoglobin gene[22] seems incompatible with the exon-domain correlation. However, Gō[35] has analysed the folding pattern of the human β-globin polypeptide, and discerns four domains. He suggests that globin genes were originally composed of four exons, and predicts an additional intron in the haem-binding exon at almost precisely the position of the extra intron found in the Lb gene. Unfortunately, other approaches for detecting globular domains in haemoglobin do not reveal these exon-related domains[36]. Thus, the correlation between exons and structural or functional domains in globins and other proteins (see Jeffreys[12]) remains highly suggestive but not completely proven.

The likely relation between exons and stable domains in proteins lends support to the idea that exons were originally 'mini-genes' that evolved to specify some simple protein function such as haem-binding[37-39]. A more sophisticated globin polypeptide could then be evolved by the rearrangement of unrelated exons and their inclusion within a common transcriptional unit; the final fusion of the exon-specified polypeptides would be achieved by RNA splicing. Similarly mutations could open up new pathways of RNA splicing in a transcriptional unit, resulting in the appearance of new combinations of exons in mature mRNA. These mechanisms for shuffling exons could provide a major source of novel genetic functions during evolution. Doolittle[40] has argued that this evolution by exon shuffling is ancient, and probably predates the divergence of prokaryotes and eukaryotes.

Darnell[38] and Crick[41] have suggested that intervening sequences might have arisen by insertion of transposable and spliceable elements into originally continuous genes. Presumably, these insertions would only be tolerated at locations where a regional disturbance of the amino acid sequence would not affect protein function; these locations might tend to be at domain boundaries and could give rise to the observed correlation between exons and protein domains. However, no example of intron gain by a gene during evolution has been documented.

There are instances where at least part of the discontinuous nature of a gene has resulted from tandem duplication of a smaller less-split gene[42]. Immunoglobulin C_H genes appear to have evolved by tandem duplication of a single C exon[43-45], and ovalbumin, ovomucoid and collagen genes also show clear signs of internal reduplications[46-48]. However, there is no evidence of any structural or functional homology between globin exons, and it seems more likely that the split gene has arisen by shuffling of unrelated exons, or by insertion of transposable elements into a once-continuous gene. In either case, introns might be regarded as non-functional DNA, and have repeatedly been included within the discussions of 'selfish' or 'junk' DNA (see Doolittle and Sapienza[49], Orgel and Crick[50], Ohno[51]).

EVOLUTION OF GLOBIN GENE CLUSTERS

The arrangement of human globin genes have been studied intensively both by restriction endonuclease analysis of human DNA, and by analysis of cloned DNA fragments containing these genes. At the moment, these genes, and their homologues in other vertebrates, provide the most detailed account of the evolutionary history of a multigene family.

Human globins are coded by two unlinked clusters of genes (see Efstratiadis et al.[4], Proudfoot et al.[52]). The α-globin gene cluster on human chromosome 16 contains two ζ- and two α-globin genes scattered over 25×10^3 base-pairs and arranged in the order 5'-ζ2-ζ1-α2-α1-3'. All genes are orientated in the same direction and are separated by substantial tracts of intergenic DNA. The β-globin gene cluster on human chromosome 11 contains five active genes spread over 45×10^3 base-pairs and arranged 5'-ϵ-$^G\gamma$-$^A\gamma$-α-β-3' (see Fig. 1). Both clusters have clearly evolved by a series of tandem globin gene duplications and have at some stage become unlinked to give the separate α- and β-globin gene clusters. After duplication, various genes must have diverged both in sequence and in developmental expression to give the current organisation of developmentally regulated genes.

Only about 8% of the DNA in these clusters is used to code globin mRNA. An additional 8% makes up the globin introns, and therefore appears in precursor mRNAs. The role of the remaining 84% of cluster DNA found between genes is a complete mystery. This intergenic DNA consists of a complex mixture of single copy DNA and elements repeated both within the cluster and elsewhere in the genome[53-56]. The small homologous sequence found near the unlinked α- and β-globin genes in mouse[14] might be an example of such a dispersed repetitive element. The function of virtually all of this intergenic DNA is a complete mystery. Close to genes, one can discern small conserved elements such as the TATA box considered to be the putative promoter for RNA polymerase II (see Breathnach and Chambon[27]). Control functions further out from globin genes are tentatively suggested by the phenotypes of various deletion mutants in the human β-globin gene cluster[57,58]. The apparent scarcity of coding sequences in the α- and β-globin gene clusters, and indeed in other mammalian gene clusters, seems to reflect the DNA excess in higher eukaryotic genomes. Indeed, Orgel and Crick[50] and Ohno[51] have suggested that these extensive intergenic regions might have no function, but instead represent 'junk' DNA. If so, then one would predict that they should evolve rapidly (at about 1 to 2%/million years) and should show rapid changes in length as a result of deletions and regional duplications driven in particular by recombination between repetitive elements in these intergenic regions.

Both of these junk predictions seem not to be borne out for the β-globin gene cluster. The δ- and β-globin gene arrangement in man was shown to be conserved in great apes and Old World monkeys[59-61]. More remarkably, the arrangement of the entire β-globin gene cluster was found to be indistinguishable in man, gorilla and baboon, indicating that the arrangement of the human cluster had been fully established 20 to 40 million years ago and faithfully preserved since[62] (Fig. 1). Furthermore, by comparing restriction endonuclease cleavage maps of intergenic DNA regions in these different primate species, it is possible to estimate rates of DNA sequence divergence in this region[62]. Surprisingly, intergenic regions evolve slowly at about 0.2% per million years (compared with 0.1%/million years for replacement site substitutions and 1 to 2%/million years predicted for junk DNA). This conservation of the arrangement and sequence of intergenic DNA strongly suggests that these sequences have been substantially constrained by selection, at least during recent primate evolution. This seems to be incompatible with the motion that intergenic DNA is junk. Instead, it might be more meaningful to regard the entire cluster (and possibly regions beyond) as a single co-adapted functional supergene rather than an assembly of largely autonomous globin genes loosely arranged by evolutionary accident into a gene cluster. One physical basis for such a large functional unit might be the existence of extensive chromatin domains including such a region; regulation of gene activity might then be seen as a consequence of modulating the packing conformation of such domains[58]. While such ideas account for the evolutionary stability of cluster arrangement, they do not readily account for the apparently strong conservation of DNA sequence.

In contrast, New World monkeys and prosimians have radically different arrangements of the β-globin gene cluster, including altered numbers of genes and major shifts in the lengths of intergenic regions[62] (see Fig. 1). However, when two diverged species of lemur were compared, the arrangement, though not sequence, of the entire cluster was found to have been completely preserved. The impression one gains is

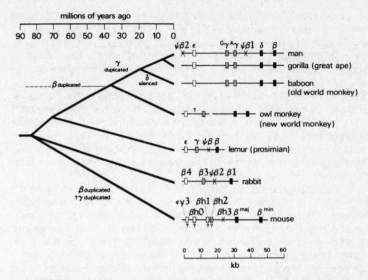

FIG. 1. *Phylogeny of the β-globin gene cluster in mammals. The arrangement of the human cluster was determined by Fritsch et al.[56], the primate clusters by Barrie et al.[62], the rabbit cluster by Lacy et al.[16] and the mouse cluster by Jahn et al.[64]. These maps show genes that are probably expressed (boxes) and known pseudogenes (crosses). All genes are transcribed from left to right, and intervening sequences within these genes are not shown. ε-Related genes are shown by open boxes, γ-related genes by hatched boxes and adult β-globin genes by filled boxes (Barrie et al.[62]). Orthologies with the mouse embryonic genes have not been determined, and the assignment of εy3 to ε and the closely related βh0 and βh1 to γ is completely speculative. Mouse βh2 and βh3 are highly divergent, and βh3, at least, is a pseudogene. Genes orthologous to human ψβ1 and ψβ2 are probably present at equivalent positions in the gorilla and baboon. The linkage of owl monkey ε- and γ-globin genes has not been firmly established, and the linkage of ε, γ to δ, β is unknown. These maps are arranged phylogenetically, using divergence times cited by Dayhoff[2], Romero-Herrera et al.[84] and Sarich and Cronin[85]. (Reprinted from Jeffreys[61]). kb, × 10³ base-pairs.*

that the arrangement of the β-globin gene cluster does not evolve smoothly, but instead proceeds in major jumps interspersed with long periods of stasis and selective constraint. It is still not certain whether such critical jumps have occurred and, if so, whether these bursts of change might be related to shifts in the physiology of erythropoiesis in primates.

Phylogenetic analyses of gene clusters, such as that shown in Fig. 1, provide a new method for timing gene duplication events. The duplicated δβ-globin gene seems to have arisen about 40 to 70 million years ago, consistent with the estimate of 40 million years by comparing δ- and β-globin gene sequences[4]. Curiously, Old World monkeys do not produce detectable δ-globin, and Martin *et al.*[59] have suggested that the gene has recently become silenced in Old World monkeys. In sharp contrast, the γ-globin gene duplicated 20 to 40 million years ago, yet the duplicates have remained almost identical in sequence in man[63]. This failure to diverge after duplication might be an example of concerted evolution (see below).

The β-globin gene cluster in the lemur (a prosimian) is the shortest so far found in mammals[62] and is similar in arrangement to the rabbit cluster[16] (see Fig. 1). This suggests that a simple cluster like that of the rabbit and lemur was established at least 85 million years ago, before the radiation of the mammals. The corresponding antiquity of the ε-, γ- and β-globin genes is consistent with their times of divergence deduced from DNA sequences[4].

The β-globin gene cluster has also been characterised in the mouse[64] (see Fig. 1). A complex cluster was found, with three to four active embryonic/foetal globin genes and duplicated adult β^{maj}- and β^{min}-globin genes. Although orthologies between the mouse non-adult genes and human εγ-globin genes have not been established, it seems likely that the mouse cluster evolved from a simple ancestral cluster by β and γ duplications independent of those seen in the lineage leading to man (Fig. 1). This model readily accounts for the reversed order of major and minor adult β-globin genes in man and mouse.

By comparison, there is little information on the evolution of the α-globin gene cluster. Zimmer *et al.*[60] compared the α2-α1-globin gene arrangement in man and apes, and found identical arrangements apart from the occasional small deletion or insertion.

Unlinked clusters of α- and β-globin genes also exist in the chicken[65,66]. However, the β-globin gene cluster contains an embryonic gene on the 3' side of the adult gene, an arrangement not seen in mammals[67]. The relation between these genes and mammalian β-related globin genes is not known: birds and mammals diverged about 270 million years ago, perhaps before the emergence of mammalian ε-, γ- and β-globin genes. It is entirely possible that the avian cluster arose by an entirely independent series of duplications of a β-related globin gene.

In sharp contrast, the major adult α1- and β1-globin genes in the amphibian *X. laevis* are closely linked in the order 5'-α1-β1-3'[68,69]. This arrangement strongly suggests that the initial αβ-globin gene duplication that occurred about 500 million years ago was a tandem duplication, and that the tandem duplicates have since remained closely linked in amphibia. In contrast, these genes probably became unlinked in the reptilian ancestors of birds and mammals, perhaps about 300 million years ago. This unlinking could have occurred by several mechanisms: inclusion of one or other globin gene within a transposable element, followed by transposition; a translocation between the α- and β-globin genes; chromosome duplication to give two unlinked αβ clusters, followed by silencing of linked β- or α-globin genes. The last model would predict the possible existence of homologous genetic functions (conserved DNA sequences) shared by regions neighbouring the mammalian α- and β-clusters; such elements should be detectable by DNA anlaysis.

$X.$ $laevis$ possesses a second $\alpha\beta$-globin gene cluster that codes for minor adult globins[68]. This cluster appears to have arisen by tetraploidisation in an ancestor of $X.$ $laevis$, and a contemporary equivalent of this ancestor, $X.$ $tropicalis$, has a single $\alpha\beta$ cluster as expected. Thus chromosome duplication has generated globin diversity in $X.$ $laevis$. The importance of chromosome duplication and polyploidisation in vertebrate evolution has been emphasised repeatedly[70-72].

Human globin gene clusters still retain the capacity for expanding and contracting their numbers of genes by unequal crossing over. The fused $\delta\beta$-globin polypeptide in haemoglobin Lepore has been shown, by direct DNA analysis, to be the product of unequal crossing-over between δ- and β-globin genes[73]. Similarly, haemoglobin Kenya probably results from unequal crossing-over between $^A\gamma$- and β-globin genes (see Jones et $al.$[74]). Chromosomes carrying three α-globin genes or a single α-globin gene have been detected in man[75-77] and the chimpanzee[60]. These three-gene and one-gene chromosomes appear to be the reciprocal products of unequal crossing-over between duplicated α-loci. The point of crossover need not necessarily be within an α-globin gene, since each α-globin gene resides within an extensive (4×10^3 base-pair) tandem repeat[15,78]. Thus, the duplicated α-locus probably arose by tandem duplication of a large region containing an α-globin gene. The mechanism for this duplication is unknown, although unequal crossing-over between sequences repeated throughout globin gene clusters[55,56] could have been responsible. By this argument, the initial spacing of gene duplicates would be dictated largely by the location of dispersed repetitive sequences relative to globin genes.

PSEUDOGENES

Recent analyses of globin gene families have revealed the existence of additional gene sequences that have accumulated mutations and no longer function as active globin genes. These additional pseudogenes have all been detected by cross-hybridisation with active globin gene sequences.

The human β-globin gene cluster contains two $\psi\beta$ sequences, one at the 5' end of the cluster and one between the $^A\gamma$- and δ-globin genes[4,56] (see Fig. 1). The sequences of these $\psi\beta$ genes have not been published, and thus their true pseudogene status is not clear. These $\psi\beta$ sequences are probably also present in apes and Old World monkeys[62]. The rabbit β-globin gene cluster contains a $\psi\beta2$ pseudogene between the $\beta3$ (γ-like) and $\beta1$(β-like) globin genes[79] (see Fig. 1). Pseudogenes are also seen (at probably similar positions) in the β-globin gene cluster of the mouse[64], lemur[62] and goat[17]. An inactive α-globin pseudogene ($\psi\gamma1$) has been found between the embryonic and adult α-globin genes in man[15]. The extraordinary mouse α-globin pseudogene, which has lost both intervening sequences[18,19], has already been described, although its linkage arrangement to functional α-globin genes has not been reported.

Several globin pseudogenes have been sequenced completely. The rabbit $\psi\beta2$ gene shows substantial divergence from the $\beta1$ gene, including frameshift mutations, premature termination codons and disruption of normal intron/exon junction sequences sufficient to render the gene incapable of coding globin[79]. The human $\psi\alpha1$ gene contains a similar range of abnormalities, plus an initiation codon mutation[15]. The mouse pseudogene βh3 (Fig. 1) shows substantial divergence from the adult β-globin gene only at the 5' end of the gene[64]. The lemur $\psi\beta$ gene appears to contain the 3' end of a β-globin gene preceded by sequences related to the 5' end of an ϵ-globin gene, although sequence data are not available[62].

Globin pseudogenes can be regarded as supernumerary sequences generated by gene duplication and silenced by divergence. There is evidence to indicate that these duplications might have initially given rise to active globin genes, which became inactivated later in evolution. For example, Lacy and Maniatis[79] compared the rabbit $\psi\beta2$ and $\beta1$ globin genes and found that replacement sites had diverged less than silent sites, to an extent that suggested that in $\psi\beta2$ the former sites (at least) were under selective constraint for some time after duplication. By using the approximate rates of replacement site and silent site divergence in active globin genes, they estimate that the duplication arose at least 50 million years ago, and that $\psi\beta2$ was eventually silenced about 30 million years ago. In a more detailed analysis of the mouse $\psi\alpha$ gene, which has lost its intervening sequences, Miyata and Yasunaga[11] calculate that the α-$\psi\alpha3$ duplication arose at least 24 million years ago and that the $\psi\alpha3$ gene was silenced 17 million years ago, after which it diverged at nearly twice the rate of silent site substitutions in active globin genes.

This apparently rapid rate of pseudogene divergence is consistent with the notion that these sequences are junk and free from selective constraint. In this respect, it will be important to carry out a direct phylogenetic comparison of homologous pseudogenes; the $\psi\beta1$ and $\psi\beta2$ genes of man, apes and Old World monkeys would provide ideal test cases. This rapid divergence also predicts the existence of many more globin pseudogenes that are too diverged to be detected by hybridisation. In contrast, the conservation of intergenic DNA in the human β-globin gene cluster, and the consistent appearance of pseudogenes between mammalian γ- and β-globin genes, might point to some functional role. Vanin et al.[19] have suggested that pseudogenes might be involved in the control of gene expression, perhaps by diverting transcription into non-productive pathways or by encoding control RNA species. Alternatively, pseudogenes might mimic active genes in maintaining the architecture of chromatin domains involved in the regulation of gene activity. If pseudogenes and active genes can interchange sequences by recombination or gene conversion (see below), then they could be of use as generators of diversity that could ultimately appear within active genes.

CONCERTED EVOLUTION

When a gene duplicates, each duplicate locus does not necessarily diverge independently. Instead, the duplicates sometimes appear to interchange sequences by some mechanism that maintains a close sequence homology between the two loci. Zimmer *et al.*[60] have called this process 'concerted evolution'.

The human $^G\gamma$- and $^A\gamma$-globin genes provide a clear example. The γ-globin gene duplication arose about 20 to 40 million years ago[62] (see Fig. 1). In contrast, the $^G\gamma$- and $^A\gamma$-globin genes isolated from a single human chromosome show extreme homology, particularly over the 5' region of the gene[63]. Furthermore, it is likely that a duplicated restriction endonuclease cleavage site polymorphism common to both γ-globin genes has arisen by sequence interchange between these loci[80]. Several mechanisms could permit such interchanges; all involve mispairing of $^G\gamma$- and $^A\gamma$-globin genes at meiosis. Recombination could then lead to the appearance of three-γ and one-γ chromosomes, or to the interchange of sequence blocks between the paired genes, or to correction of one sequence to the other by gene conversion (see Slightom *et al.*[63], Zimmer *et al.*[60], Jeffreys[12]). In all cases, homogenisation of sequence differences could ensue, leading to concerted evolution at the duplicated γ-locus.

A similar phenomenon appears to be occurring at the duplicated $\alpha2$-$\alpha1$ locus. The duplication arose at least 8 million years ago[60], yet the $\alpha2$- and $\alpha1$-globin genes are almost identical in sequence[81]. Again, homology is most marked in a localised area of these genes, suggesting that the most recent round of 'correction' was confined to only part of the α-globin gene sequence. The ability of α-globin genes to mispair and recombine at meiosis is strongly supported by the existence of three-α and one-α loci in man (see above).

Concerted evolution by gene conversion or multiple unequal crossing-over can also be invoked to account for the lack of divergence of the first intervening sequence in the mouse β^{maj}- and β^{min}-globin genes[25], the relative similarity of the 5' ends of the human β- and δ-globin genes[4], the remarkable homology of 5' flanking regions near the goat γ- β^c- and β^A-globin genes[82], and what appears to be an ϵ–β hybrid structure of the lemur $\psi\beta$ gene[62]. Clearly, concerted evolution is not a rare phenomenon, and seems to occur between even relatively distantly related genes and between active genes and pseudogenes. One wonders about the level of selective constraint required to prevent this type of process from completely homogenising a globin gene cluster.

Concerted evolution is not restricted to globin genes, and has been documented in a variety of other gene families and in repetitive DNA sequences (see Dover and Coen[83], Jeffreys[12]).

CONCLUDING REMARKS

The application of recombinant DNA methods to the study of gene evolution, especially of globin genes, is rapidly producing a mass of data concerning rates and modes of gene evolution, and the types of informational interchange and selective constraints that might operate on gene families and clusters. While much more information is required on functional gene sequences, the time is ripe for focusing attention on clusters and, particularly, intergenic DNA. How much of this intergenic DNA is occupied by divergent pseudogenes? Are they really functionless? Can one discern large conserved domains encompassing entire clusters? Do clusters evolve in a discontinuous fashion? Do transposable elements exist in higher eukaryotes and what influence would they have on the appearance and dispersal of multigene families?

As yet, these studies give few clues as to key events responsible for karyotypic evolution, speciation, and morphological and behavioural shifts in evolution. Yet they are important for understanding the basic rules that govern DNA evolution, and time alone should enable us to bridge the gap between these fine structural studies of molecular evolution and the effects of these molecular changes on phenotype and their relation to macro-evolutionary phenomena.

1. Hunt, T. L., Hurst-Calderone, S. & Dayhoff, M. O. In 'Atlas of protein sequence and structure' vol. 5 suppl. 3, 229–251 (1978).
2. Dayhoff, M. O. Editor of Atlas of protein sequence and structure, vol. 5, Natl. Biomed. Res. Found, Silver Spring, Md. (1972).
3. Wilson, A. C., Carlson, S. S. & White, T. J. Annu. Rev. Biochem. 46, 573–639 (1977).
4. Efstratiadis, A., Posakony, J. W., Maniatis, T., Lawn, R. M., O'Connell, C., Spritz, R. A., DeRiel, J. K., Forget, B. G., Weissman, S. M., Slightom, J. L., Blechl, A. E., Smithies, O., Baralle, F. E., Shoulders, C. C. & Proudfoot, N. J. Cell 21, 653–668 (1980).
5. Kafatos, F. C., Efstratiadis, A., Forget, B. G. & Weissman, S. M. Proc. Nat. Acad. Sci., U.S.A. 74, 5618–5622 (1977).
6. Perler, F., Efstratiadis, A., Lomedico, P., Gilbert, W., Kolodner, R. & Dodgson, J. Cell 20, 555–566 (1980).
7. Jukes, T. H. Science 210, 973–978 (1980).
8. Jukes, T. H. & King, J. L. Nature (London) 281, 605–606 (1979).
9. Miyata, T., Yasunaga, T. & Nishida, T. Proc. Nat. Acad. Sci. 77, 7328–7332 (1980).
10. Kimura, M. Proc. Nat. Acad. Sci., U.S.A. 78, 454–458 (1981).
11. Miyata, T. & Yasunaga, T. Proc. Nat. Acad. Sci., U.S.A. 78, 450–453 (1981).
12. Jeffreys, A. J. In 'Genetic Engineering'. (Williamson, R., ed.), vol. 2, Academic Press, London, New York (1981).
13. Grantham, R., Gautier, C., Gouy, M., Jacobzone, M. & Mercier, R. Nucl. Acids Res. 9, r43–r74 (1981).
14. Leder, A., Miller, H. I., Hamer, D. H., Seidman, J. G., Norman, B., Sullivan, M. & Leder, P. Proc. Nat. Acad. Sci., U.S.A. 75, 6187–6191 (1978).
15. Proudfoot, N. J. & Maniatis, T. Cell 21, 537–545 (1980).
16. Lacy, E., Hardison, R. C., Quon, D. & Maniatis, T. Cell 18, 1273–1283 (1979).
17. Cleary, M. L., Haynes, J. R., Schon, E. A. & Lingrel, J. B. Nucl. Acids Res. 8, 4791–4802 (1980).
18. Nishioka, Y., Leder, A. & Leder, P. Proc. Nat. Acad. Sci., U.S.A. 77, 2806–2809 (1980).
19. Vanin, E. F., Goldberg, G. I., Tucker, P. W. & Smithies, O. Nature (London) 286, 222–226 (1980).
20. Lomedico, P., Rosenthal, N., Efstratiadis, A., Gilbert, W., Kolodner, R. & Tizard, R. Cell 18, 545–558 (1979).
21. Bell, G. I., Pictet, R. L., Rutter, W. J., Cordell, B., Tischer, E. & Goodman, H. M. Nature (London) 284, 26–32 (1980).
22. Jensen, E. O., Paludan, K., Hyldig-Nielsen, J. J., Jorgensen, P. & Marcker, K. A. Nature (London) 291, 677–679 (1981).
23. Goff, S. P., Gilboa, E., Witte, O. N. & Baltimore, D. Cell 22, 777–785 (1980).
24. Van Den Berg, J., Van Ooyen, A., Mantei, N., Schambӧck, A., Grosveld, G., Flavell, R. A. & Weissmann, C. Nature (London) 276, 37–44 (1978).
25. Konkel, D. A., Maizel, J. V. & Leder, P. Cell 18, 865–873 (1979).
26. Van Ooyen, A., Van Den Berg, J., Mantei, N. & Weissmann, C. Science 206, 337–344 (1979).
27. Breathnach, R. & Chambon, P. Annu. Rev. Biochem. 50, in the press (1981).
28. Wahli, W., Dawid, I. B., Wyler, T., Weber, R. & Ryffel, G. U. Cell 20, 107–117 (1980).
29. Jones, R. E., Bhat, S. P., Sullivan, M. A. & Piatigorsky, J. Proc. Nat. Acad. Sci., U.S.A. 77, 5879–5883 (1980).
30. Heilig, R., Perrin, F., Gannon, F., Mandel, J. L. & Chambon, P. Cell 625–637 (1980).
31. Blake, C. C. F. Nature (London) 277, 598 (1979).
32. Craik, C. S., Buchman, S. R. & Beychok, S. Proc. Nat. Acad. Sci., U.S.A. 77, 1384–1388 (1980).
33. Craik, C. S., Buchman, S. R. & Beychok, S. Nature (London) 291, 87–90 (1981).
34. Eaton, W. A. Nature (London) 284, 183–185 (1980).
35. Gō, M. Nature (London) 291, 90–92 (1981).
36. Rashin, A. A. Nature (London) 291, 85–87 (1981).
37. Gilbert, W. Nature (London) 271, 501 (1978).
38. Darnell, J. E. Science 202, 1257–1260 (1978).
39. Reanney, D. Nature (London) 277, 598–600 (1979).
40. Doolittle, W. F. Nature (London) 272, 581–582 (1978).

41. Crick, F. *Science* **204**, 246–271 (1979).

42. Ohno, S. 'Evolution by gene duplication'. Springer-Verlag, Berlin, Heidelberg & New York (1980).

43. Early, P. W., Davis, M. M., Kaback, D. B., Davidson, N. & Hood, L. *Proc. Nat. Acad. Sci., U.S.A.* **76**, 857–861 (1979).

44. Sakano, H., Rogers, J. H., Hüppi, K., Brack, C., Trannecker, A., Maki, R., Wall, R. & Tonegawa, S. *Nature (London)* **277**, 627–633 (1979).

45. Sakano, H., Hüppi, K., Heinrich, G. & Tonegawa, S. *Nature (London)* **280**, 288–294 (1979).

46. Cochet, M., Gannon, F., Hen, R., Maroteaux, L., Perrin, F. & Chambon, P. *Nature (London)* **282**, 567–574 (1979).

47. Stein, J. P., Catterall, J. F., Kristo, P., Means, A. R. & O'Malley, B. W. *Cell* **21**, 681–687 (1980).

48. Ohkubo, H., Vogeli, G., Mudryj, M., Avvedimento, V. E., Sullivan, M., Pastan, I. & De Grombrugghe, B. *Proc. Nat. Acad. Sci., U.S.A.* **77**, 7059–7063 (1980).

49. Doolittle, W. F. & Sapienza, C. *Nature (London)* **284**, 601–603 (1980).

50. Orgel, L. E. & Crick, F. H. C. *Nature (London)* **284**, 604–606 (1980).

51. Ohno, S. *Rev. Brasil. Genet. III*, **2**, 99–114 (1980).

52. Proudfoot, N. J., Shader, M. H. M., Manley, J. L., Gefter, M. L. & Maniatis, T. *Science* **209**, 1329–1336 (1980).

53. Adams, J. W., Kaufman, R. E., Kretschmer, P. J., Harrison, M. & Nienhuis, A. W. *Nucl. Acids Res.* **8**, 6113–6128 (1980).

54. Baralle, F. E., Shoulders, C. C., Goodbourn, S., Jeffreys, A. & Proudfoot, N. J. *Nucl. Acids Res.* **8**, 4393–4404 (1980).

55. Coggins, L. W., Grindlay, G. J., Vass, J. K., Slater, A. A., Montague, P., Stinson, M. A. & Paul, J. *Nucl. Acids Res.* **8**, 3319–3333 (1980).

56. Fritsch, E. F., Lawn, R. M. & Maniatis, T. *Cell* **19**, 959–972 (1980).

57. Fritsch, E. F., Lawn, R. M. & Maniatis, T. *Nature (London)* **279**, 598–603 (1979).

58. Van Der Ploeg, L. H. T., Konings, A., Oort, M., Roos, D., Bernini, L. & Flavell, R. A. *Nature (London)* **283**, 637–642 (1980).

59. Martin, S. L., Zimmer, E. A., Kan, Y. W. & Wilson, A. C. (1980). *Proc. Nat. Acad. Sci., U.S.A.* **77**, 3563–3566 (1980).

60. Zimmer, E. A., Martin, S. L., Beverley, S. M., Kan, Y. W. & Wilson, A. C. *Proc. Nat. Acad. Sci. U.S.A.* **77**, 2158–2162 (1980).

61. Jeffreys, A. J. & Barrie, P. A. *Phil. Trans. Roy. Soc. Ser. B* **292**, 133–142 (1981).

62. Barrie, P. A., Jeffreys, A. J. & Scott, A. F. *J. Mol. Biol.* **149**, 319–336 (1981).

63. Slightom, J. L., Blechl, A. E. & Smithies, O. *Cell* **21**, 627–638 (1980).

64. Jahn, C. L., Hutchison, C. A., Phillips, S. J., Weaver, S., Haigwood, N. L., Voliva, C. F. & Edgell, M. H. *Cell* **21**, 159–168 (1980).

65. Hughes, S. H., Stubblefield, E., Payvar, F., Engel, J. D., Dodgson, J. B., Spector, D., Cordell, B., Schimke, R. T. & Varmus, H. E. *Proc. Nat. Acad. Sci., U.S.A.* **76**, 1348–1352 (1979).

66. Engel, J. D. & Dodgson, J. B. *Proc. Nat. Acad. Sci., U.S.A.* **77**, 2596–2600. (1980).

67. Dodgson, J. B., Strommer, J. & Engel, J. D. *Cell* **17**, 879–887 (1979).

68. Jeffreys, A. J., Wilson, V., Wood, D., Simons, J. P., Kay, R. M. & Williams, J. G. *Cell* **21**, 555–564 (1980).

69. Patient, R. K., Elkington, J. A., Kay, R. M. & Williams, J. G. *Cell* **21**, 565–573 (1980).

70. Ohno, S. *Nature (London)* **244**, 259–262 (1973).

71. Lalley, P. A., Minna, J. D. & Francke, U. *Nature (London)* **274**, 160–162 (1978).

72. Lundin, L.-G. *Clin. Genet.* **16**, 72–81 (1979).

73. Flavell, R. A., Kooter, J. M., De Boer, E., Little, P. F. R. & Williamson, R. *Cell* **15**, 25–41 (1978).

74. Jones, R. W., Old, J. M., Trent, R. J., Clegg, J. B. & Weatherall, D. J. *Nature (London)* **291**, 39–44 (1981).

75. Embury, S. H., Miller, J., Chan, V., Todd, D., Dozy, A. M. & Kan, Y. W. *Blood* **54** (suppl.), 53a (1979).

76. Goossens, M., Dozy, A. M., Embury, S. H., Zachariades, Z., Hadjiminas, M. G., Stamatoyannopoulos, G. & Kan, Y. T. *Proc. Nat. Acad. Sci., U.S.A.* **77**, 518–521 (1980).

77. Higgs, D. R., Old, J. M., Pressley, L., Clegg, J. B. & Weatherall, D. J. *Nature (London)* **284**, 632–635 (1980).

78. Lauer, J., Shen, C. K. J. & Maniatis, T. *Cell* **20**, 119–130 (1980).

79. Lacy, E. & Maniatis, T. *Cell* **21**, 545–553 (1980).

80. Jeffreys, A. J. *Cell* **18**, 1–10 (1979).

81. Liebhaber, S. A., Goossens, M. & Kan, Y. W. *Nature (London)* **290**, 26–29 (1981).

82. Haynes, J. R., Rosteck, P. & Lingrel, J. B. *Proc. Nat. Acad. Sci., U.S.A.* **77**, 7127–7131 (1980).

83. Dover, G. & Coen, E. *Nature (London)* **290**, 731–732 (1981).

84. Romero-Herrera, A. E., Lehmann, H., Joysey, K. A. & Friday, A. E. *Nature (London)* **246**, 389–395 (1973).

85. Sarich, V. M. & Cronin, J. E. *Nature (London)* **269**, 354 (1977).

A. J. Jeffreys is in the Genetics Department, University of Leicester.

This article first appeared in *Genome evolution* (eds. Dover, G. A. & Flavell, R. B.) pp. 157–176 (Academic Press, 1982). It is reproduced with permission.

CHAPTER THREE

LAMARCKIAN INHERITANCE AND THE PUZZLE OF IMMUNITY

NEO-DARWINISM is essentially Weismannist; that is, it holds that heritable variation is in origin non-adaptive, and that the adaptive changes which occur during the lifetime of individuals (for example, acclimatisation to high altitudes) do not alter the nature of the offspring they produce. In contrast, Darwin accepted a Lamarckian view of heredity; he thought that the 'effects of use and disuse' could influence the nature of offspring, and even developed his theory of pangenesis to account for such effects. When he said that he rejected Lamarckism, it was not this theory of heredity he was rejecting, but the idea that evolution could be explained by an inner drive towards complexity; quite rightly, he saw that this is an explanation which explains nothing.

If Weismann was right, then natural selection plays an even more fundamental role than Darwin thought; it becomes the only process leading to genetic adaptation. Today, our main reason for being Weismannist is that most innate differences between similar organisms appear to be caused by differences between nucleic acids, and there are good reasons to accept the 'central dogma' of molecular biology, that information does not pass from proteins to nucleic acids. However, the question of the origin of hereditary variation remains central to evolutionary biology, if only because Lamarck's theory is the only alternative to Darwin's that has been suggested.

I have thought for many years that anyone with Lamarckian leanings should look to immunology for confirmation. At least superficially, the facts concerning immunity suggest a Lamarckian interpretation. Thus a vertebrate can become immune to a foreign substance which neither it nor its ancestors have ever met. This means that it 'learns' to produce a specific protein—an antibody—able to bind to that substance. Since proteins are coded for by DNA and RNA, the phenomenon of immunity suggests that information has somehow passed from the 'environment'—the foreign substance—to RNA and/or DNA. The phenomenon of tolerance, whereby individuals learn not to produce antibodies against their own tissues, points to a similar conclusion.

Nevertheless, immunologists have been reluctant to accept this Lamarckian interpretation. Instead, they have sought mechanisms whereby the body could produce a great variety of cells, each capable of producing a different antibody—the 'generation of diversity', or GOD—so that selection between the cells of the immune system could then favour the multiplication of that cell which, by chance, happened to produce the appropriate antibody. It now looks as if this hunch was correct; we are close to understanding how GOD works. The article by Miranda Robertson describes the present state of play. If, as now seems certain, a neo-Weismannist rather than a neo-Lamarckian interpretation of immunity proves to be correct, it will confirm the wisdom of not abandoning a well-established theory at the first sign of factual contradiction. Not every anomaly is the starting point of a new paradigm.

The last few years has seen a curious twist to this story in the work of E. J. Steele and R. M. Gorczynski, commented on here by Jonathan Howard. Steele, in his book *Somatic Selection and Adaptive Evolution*, suggested the following mechanism whereby acquired characters might be transmitted to offspring. He supposes that variation between somatic cells is generated by some kind of mutational process, and that developmental adaptation occurs because of the selective multiplication of some kinds of cells; as far as the immune system is concerned this supposition is reasonable, although it is hard to see how it could apply to most other developmental adaptations. Steele then suggests that the DNA-coded information from the selected cells is carried by RNA viruses to the germ cells and there incorporated into the chromosomes, resulting in the transmission of the acquired adaptation to sexually-produced offspring. Note that Steele's mechanism does not require that information pass from protein to nucleic acid—it does not contravene the central dogma. It does require that information pass from RNA to DNA, a process which is known to occur.

Steele's process would mimic Lamarckian inheritance, without being Lamarckian at the molecular level. Oddly enough, it is remarkably similar to the process of 'tissue selection' proposed by Weismann, although Weismann seems to have been more aware of its limitations: essentially, these are that most developmental adaptations do not involve the multiplication of specific nucleic acid molecules, and that some non-adaptive changes (cancers) do.

Howard describes the reception of Steele's ideas, Gorczynski's experimental tests, and some unsuccessful attempts to replicate them. It is right that the idea should have been listened to, and that serious attempts should have been made to repeat the experiments. It is less obvious that a major British newspaper was right to announce, in a centre page spread based on Steele's work, that Darwinism is dead.

Genes of lymphocytes: Diverse means to antibody diversity

MIRANDA ROBERTSON

AFTER fifteen years of controversy and five years of DNA cloning, immunologists are at last unanimous on how antibody diversity is generated. This remarkable eventuality is due to the accumulation over the past five years of evidence for almost every mechanism anyone has ever suggested; and to which has now been added yet another, of such baroque complexity that it could never have occurred to anyone to suggest it but for the evidence (albeit preliminary) in the DNA.

There used to be, broadly, three schools of thought, one of which adopted the extreme position that the entire antibody repertoire was encoded in a vast array of germ-line genes, and the other two of which favoured the somatic generation of variation on the basis of a limited number of germ-line genes. Two means were proposed for the somatic generation of variation. One was simple point mutation; but this on its own would not explain the distribution of the variation, which is concentrated in three 'hypervariable' regions on the variable region gene (see Fig. 1). The other was recombination, represented in its most elegant and extreme form by the mini-gene theory of Kabat, who proposed that a range of genes coding for each of the three hypervariable regions and the four less variable (framework) regions existed separately in the genome, to be brought together in variable combinations during the differentiation of the B lymphocyte. This theory however did not explain the evidence for point mutations in the framework regions. All these views have now been at least partly vindicated, and their proponents are still reeling with surprise at finding they no longer have anything to quarrel about.

Two things became clear very soon after the cloning and sequencing of the first immunoglobulin genes. First, there is quite a large number of germ-line genes: more than one for each of the known variable region subgroups (a subgroup is a family of variable regions defined by similarity in the framework regions). There are certainly some hundreds, and there may be some thousands of germ-line V genes.

Second, the differentiation of a B lymphocyte is accompanied by the translocation of V region DNA. In the case of the light chain, one of the germ-line V genes, comprising the first three framework segments with

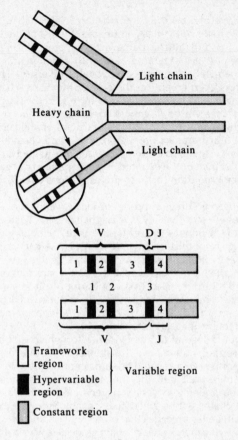

Fig. 1. *Immunoglobulin molecules are made up of four chains: one pair of heavy chains and one pair of light chains. Each chain is divided functionally into two regions: the variable region, which mediates antigen recognition, and the constant region, which in the case of the heavy chains mediates the effector function of the molecule.*

The variable region can be further subdivided into four regions which vary only slightly from one antibody molecule to another, and are known as the framework regions, and three regions (the hypervariable regions) which are very much more variable and in the folded protein form the actual antigen-binding site.

Each light chain is now known to be coded by three separate genes: a C gene coding for the constant region, a J gene coding for the fourth framework region and part of the third hypervariable region, and a V gene coding for the first three framework regions including hypervariable regions one and two and part of hypervariable region three.

Each heavy chain is coded by four separate genes: a C gene coding for the constant region, a J gene coding for the fourth framework region, a D gene coding for the third hypervariable region, and a V gene coding for the first three framework regions including hypervariable regions one and two.

two hypervariable regions and part of the third, is brought into proximity with one of a cluster of J genes comprising the rest of the third hyper-variable region and the fourth framework segment. Since V genes may recombine with J genes at random, and with variation in the site of the join, this immediately generates considerable diversity. The scope for diversification is even greater in the case of the heavy-chain genes, in which the V gene, which extends only to the end of the third framework region, recombines with a D gene comprising the third hypervariable region, and the D in turn recombines with one of a cluster of J genes coding for the fourth framework region. The D and J heavy-chain genes thus behave very much like the mini-genes postulated by Kabat; but Kabat never dreamed of the extraordinary properties of D genes that are beginning to emerge from the latest investigations of S. Tonegawa and his colleagues.

They have used a D region probe[1] to locate a cluster of eight homo-logous D genes in germ-line DNA. All eight genes are 17 base-pairs long; yet expressed D regions with substantial 'core' homology to the germ-line genes range from thirteen to more than 40 base-pairs in length. This is too extreme to be explained by variations in the site of V-D-J re-combination. Tonegawa has suggested that the explanation may lie in D-D joining, by a mechanism that would not result in simple tandem repetition of the homologous sequences. The mechanism depends on the 12/23 rule first enunciated by Early et al.[2] for V-J recombination, and is described in detail in the legend to Fig. 2.

Even without D-D joining, the evidence for which is at best circum-stantial, the recombination of a substantial repertoire of V genes at random with J and D genes would be sufficient to account for an antigen-binding repertoire well into the thousands. To this variation within chains must be added the variation generated by the association of different light chains with a given heavy chain. Since the expression of the heavy-chain genes precedes that of the light-chain genes, it has been assumed that each different heavy chain associates with a different light chain in the progeny of a single pre-B cell[4]. By fusing a pre-B cell line with a myeloma line that does not express L chains, Kuehl and his colleagues have succeeded in inducing the expression of light chain genes of the pre-B cell and demonstrated that different rearrangements of the κ genes do in fact occur in the progeny[5].

Finally, there is growing evidence for further diversification of the rearranged immunoglobulin genes by somatic mutation in the course of B cell ontogeny. In the absence of a B cell line whose ontogeny can be followed *in vitro*, the evidence is necessarily indirect. But the availability of cloned germ-line genes for the V-regions of some antibodies has made it possible to compare the germ-line sequences with those of secreted immunoglobulins from B cells at different stages of maturation.

The different stages of B cell maturation are marked by a switch in the constant region of the antibody, which determines its class. Early in their ontogeny, B cells synthesize immunoglobulin M; at later stages they synthesize immunoglobulins of other classes: IgG, IgA or IgE. Thus by comparing IgM and (say) IgG in turn with the homologous germ-line

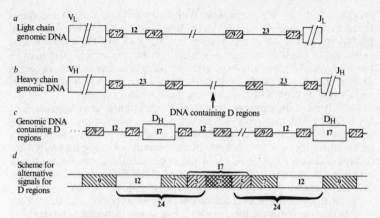

FIG. 2. *The pattern of genomic sequences believed to mediate V-J and V-D-J joining in light (a) and heavy (b and c) chain variable region genes. Open rectangles are coding sequences, shaded rectangles are nonamer and heptamer signal sequences, and intervening lines are non-coding DNA. The figures 12 and 23 on the non-coding DNA refer to the length of the spacer regions between the heptamer and the nonamer, which are highly conserved sequences believed to act as recombination signals: the heptamer is followed by the nonamer after a space of 12 (strictly, 11 or 12) base pairs downstream of the V_L gene, and inversely, the nonamer is followed by the heptamer after a space of 23 (strictly, 22 or 24) base pairs upstream of the J_L gene (Fig. 2 a). Heavy-chain V genes are followed by the same signal sequences with the 23 spacer, and heavy-chain J genes are preceded by the inverse signal, again with a 23 spacer (Fig. 2b). Early et al.[2] and Sakano et al.[3] surmised that recombination depends on the pairing of the signal sequences plus a 12-base-pair spacer with the inverse signal sequences plus a 23-base-pair spacer; they accordingly predicted that D_H genes should be flanked by signals with 12-base-pair spacers (Fig. 2 c). This has proved to be the case[2]. If the 12/23 rule holds, however, it should make D-D joining impossible; and that is where the baroque complexity comes in.*

Tonegawa and Kurosawa have proposed that each D region can treat the central 7 of its 17 base-pairs as the heptamer of the signal sequences and add the five coding base-pairs plus their adjoining heptamer, either upstream of downstream, to the adjoining spacer to make a 24 instead of a 12 base-pair spacer (Fig. 2 d). This would make it possible to add about 5 base-pairs of one D gene to another without violating the 12/23 (or 12/24) rule or producing tandem repeats, which are not found.

sequence, it is possible to detect any increase in departures from the germ-line gene that may have occurred as the cells switched from one class of immunoglobulin to the next.

Such increases have already been reported by Bothwell *et al.*[6], working with antibodies directed against nitrophenol (NP), and by Gearhart *et al.*[7], working with antibodies against phosphorylcholine. Antibodies raised against these antigens possess, in certain strains of mouse, the interesting and useful property that a very high proportion of the total immunoglobulin bears a characteristic antigenic determinant, or idiotype, of its own (NP[b] in the case of anti-NP antibodies, and T15 in the case of anti-PC antibodies). But not all even of those antibodies bearing the same idiotype are by any means identical, and in the light of recent analyses this is hardly surprising. Both Bothwell[6] and Gearhart find that there is a cluster of several germ-line V genes all with substantial homology to the idiotype-bearing anti-NP and anti-PC proteins. The existence of such homologous V-gene clusters (as well as Tonegawa's homologous D gene clusters) strongly suggests duplication of genes during evolution—particularly since two of the five homologous genes sequenced by Gearhart et al. are actually identical.

As it turns out, comparison of the partial amino-acid sequences of several T15 proteins with the germ-line sequences shows that all the proteins are derived from the two identical germ-line genes, but with numerous substitutions, notably in the proteins of ontogenetically later classes (IgG and IgA)[7]. None of these substitutions can be explained except by mutation, with one exception which could (but need not) be due to recombination between two of the germ-line V genes.

The sum of all these recent data leaves us with an embarrassing riches of diversity-generating mechanisms, but some important outstanding questions about their deployment. The first and most obvious is exactly how many germ-line genes there are. It is believed that there are four functional J genes each for the light and heavy chains; but estimates of the numbers of V genes still range between 500 and 1,000, and little is known of the total size of the D-gene family, of which that sequenced by Tonegawa and his associates may represent only a branch.

The second and more challenging is whether a special mechanism, still undiscovered, is needed to explain the concentration of amino-acid substitutions in the hypervariable regions. Both germ-line sequence and somatic recombination go some way towards explaining this pattern of variation. Related germ-line genes differ from one another more in the hypervariable regions than in framework-coding regions; and variable V-J and V-D-J joining account for further variation in the third hypervariable region—particularly in the case of V-D-J joining. But Gearhart et al.[7] have shown that somatic mutations too are clustered, at hypervariable regions 1 and 2.

This could in principle be explained without recourse to further special mechanisms, simply by somatic selection. B cells whose surface immunoglobulin does not bind antigen produce no progeny: thus those expressing mutations in the hypervariable regions which form the antigen-binding site will tend to be selected. Selection will also occasionally operate on mutations affecting framework sequences: although it is the hypervariable regions that actually make contact with antigen, the framework regions may influence the manner in which they do so—for example, framework variants of an immunoglobulin which in its original form bound PC both on its own and complexed to a carrier will no longer bind the PC-carrier complex[8]. Both the clustered substitutions in the hypervariable regions, and the occasional ones in the framework can therefore be accounted for by selection for antigen binding.

It would be a mistake however to assume that all mutations in expressed immunoglobulin genes must have been specifically selected: indeed, some are silent and could never have been subjected to selection. These are presumably the 'neutral mutations' of the immune system. But if the pattern of mutations requires no special explanation, it is possible that the rate of mutation may. Certainly many immunologists are wondering about possible mutagenic reactions associated with the class switch. Notwithstanding the considerable difficulties associated with the elucidation of molecular mechanisms of immunoglobulin gene re-arrangement, the issue of mutation rate may be resolved more quickly by research at this level than by arguments about sequences, because arguments about the mutation and selection of sequences depend on a detailed understanding of the relationship between the sequence of an immunoglobulin and its antigen binding properties and this is usually unknown.

However, each antibody molecule can now be seen as the end product of a series of selective steps, beginning with the germ-line genes. The germ-line V, J and D genes have presumably been selected in the course of evolution for their ability to produce proteins that will bind to common environmental antigens. There follow at least two rounds of somatic selection by antigen. The first acts on the V-J and V-D-J recombinations that produce IgM. It is followed by at least one more round of somatic selection favouring mutations that produce efficient binding to the same antigen, or perhaps to other antigens present in large enough quantities to encounter the mutant cell. A detailed analysis of hybridoma proteins from immunized and hyperimmunized mice, such as Bothwell and his colleagues have already begun, may help to show how far antigenic selection could account for the range of antibodies a single mouse is able to produce against a single antigen.

1. Sakano, H., Kurosawa, Y., Weigert, M. & Tonegawa, S. Nature 290, 562 (1981).
2. Early, P. et al., Cell 19, 981 (1980).
3. Sakano, H. et al., Nature 286, 676 (1980).
4. Burrows, P., Lejeune, M. & Kearney, J. F. Nature 280, 838 (1979).
5. Riley, S. C., Brock, E. J. & Kuehl, W. M. Nature 289, 804 (1981).
6. Bothwell, A. L. M., et al., Cell 24, 625 (1981).
7. Gearhart, P., Johnson, N. D., Douglas, R. & Hood, L. E. Nature 291, 29 (1981).
8. Scharff, M. D. Proc. natn. Acad. Sci. U.S.A. (in the press).

Miranda Robertson is associate editor of Nature.

This article first appeared in *Nature* Vol. **290**, pp. 625–627; 1981.

A tropical Volute shell and the Icarus syndrome

JONATHAN C. HOWARD

DARWIN tells a little parable in his autobiography[1] which he learned as a very young man, before the *Beagle* voyage, from Adam Sedgwick, the Professor of Geology in Cambridge.

'Whilst examining an old gravel-pit near Shrewsbury, a labourer told me that he had found in it a large worn tropical Volute shell, such as may be seen on the chimney pieces of cottages; and as he would not sell the shell, I was convinced he had really found it in the pit. I told Sedgwick of the fact, and he at once said (no doubt truly) that it must have been thrown away by some one into the pit; but then added, if really embedded there it would be the greatest misfortune to geology as it would overthrow all that we know about the superficial deposits of the Midland Counties. These gravel-beds belong in fact to the glacial period, and in after years I found in them broken Arctic shells. But then I was utterly astonished at Sedgwick not being delighted at so wonderful a fact as a tropical shell being found near the surface in the middle of England.'

The point of the parable is that science does not merely consist of making observations but also of ordering them in relation to each other. The tropical shell did not fit into an ordered scheme, and so either the shell or the scheme was wrong. Sedgwick backed the scheme against the shell and won. There would be nothing unscientific about backing the shell, but the quality of support would have to be strong in proportion to the length of the odds against it. No stone could be left unturned to verify every aspect of the find, and eventually a hypothesis would have to be formulated of sufficient generality to accommodate the coexistence of Arctic and tropical shells in the same deposits. Even then the pre-existing scheme would probably not be 'wrong' but merely incomplete. Sedgwick rather overstated the case. New, confirmed and revolutionary discoveries in science very rarely utterly degrade an existing scheme and rebuild its elements in an entirely new pattern. In biology only the theory of evolution by natural selection came close to doing this, and most of the material incorporated into the new theory was altered only in context rather than form.

The scientific temperament varies, no doubt, along many dimensions. But certainly one of the most interesting is the dimension of credulity.

How eccentric does a finding or proposition have to be before it runs off the end of your credulity scale? Are you for the shell or for Sedgwick? Incredulity was my first reaction when I heard by the usual roundabout route that Dr Gorczynski of the Ontario Cancer Institute was reporting the transmission of acquired immunological tolerance through the male. I assumed that the grapevine, with its usual capacity for decreasing the signal-to-noise ratio, had got the sex wrong. Maternal transmission would have fitted comfortably into a well ordered scheme since mother and fetus are known to engage in a rather wide-ranging immunological discourse both before and after parturition; but the grapevine had got the sex right, and here was our tropical Volute shell in the glacial gravel pit. What the grapevine did not say was anything about the genesis of the experiment, and this was a grave omission. Gorczynski's collaboration with Dr E. Steele was conducted with the specific intention of searching for paternal transmission of tolerance, under the influence of Steele's already formulated hypothesis[2] that RNA viruses might act as vectors for the horizontal transmission of the DNA-coded information from somatic cells to cells of the germ line. According to this hypothesis, transmission through the male is more probable than through the female because male germ cells are formed continuously throughout life.

The phenomenon became public knowledge when the experiments were first published last year[3]. The level at which I have heard these experiments and their sequel[4] discussed has varied from gleeful iconoclasm, through critical scepticism tinged with curiosity, to frank disbelief. More upsetting is the charge that the data are suspect because the experiments were designed and conducted under the influence of a hypothesis of high import. Under these conditions, so the argument runs, the eye is guided more by hope than by the ineluctable evidence of the γ-counter. I assumed that this naïve folly had been killed by Sir Peter Medawar's graceful and persistent advocacy of the importance of hypothesis, as a guide to useful scientific action, but under the provocation of Gorczynski and Steele's results it seems to have risen from the grave. Since Darwin is often pushed into the ring to confront Steele, I might add that Darwin unequivocally endorsed Steele's logistical procedures when he wrote[5]: 'How odd it is that anyone should not see that all observation must be for or against some view if it is to be of any service'.

As for the gleeful iconoclasm, it reminds me of the scornful laughter which greeted my Latin master when he muffed an easy declension. He remained an incomparably better classicist than we were. How much of neo-Darwinism are the iconoclasts prepared to throw out on the strength of Gorczynski and Steele's findings? Weissmann's doctrine of the isolation of the germ plasm from somatic influence was a brilliant hypothesis. If it had been badly wrong, genes would not behave in the elegant and predictable way that they do. In the context of Steele's hypothesis, if viral vectors are engaged in the transmission of DNA between cells, and, as implied by Gorczynski and Steele's recent paper in *Nature*[4], are no respecters of the linkage group, we should have no loci and no alleles. The chromosomes of the germ cells would be a randomized dump of DNA sequences gathered from cells all over the body (see Darwin's

Provisional Hypothesis of Pangenesis for a nineteenth century expression of this idea)[6]. Every time alleles behave like alleles and linkage groups behave like linkage groups it is another confirmation of the generality of mendelian genetics and of the integrity of the germ line, and a further exception to the generality of Steele's hypothesis. However, since neither Weissmann's doctrine nor mendelian genetics predict Gorczynski and Steele's findings, the obligation falls on everyone concerned to find out what kind of exception to the order of things these phenomena are.

It is the mixture of critical scepticism and curiosity alone that is going to resolve this matter. It is a tribute to the vigour of these faculties in the immunological community that already *Nature* has carried two manuscripts which go directly to the heart of the scientific problems on hand. First and foremost, are Gorczynski and Steele's results readily reproducible? Brent and colleagues[7] have conducted one of the most serious attempts I have ever seen in immunology to reproduce an experimental protocol in detail. Furthermore, the analysis of the phenomenon is extended in two ways: tolerance was assessed by skin graft rejection as well as the capacity to generate cytotoxic cells *in vitro*, and repeated tests of *in vitro* reactivity were performed on individual animals. There is no escaping the conclusion that, under the conditions of these experiments, the Gorczynski and Steele result has not been confirmed. The explanation for the inconsistency between the two results is now confined to the residual procedural differences between the two series of experiments. Brent and colleagues draw attention to several such differences, only one of which is, in my opinion, important. This difference is in the cell populations used to induce tolerance in the experimental fathers. Gorczynski and Steele's tolerance-inducing protocol was eccentric in the extreme: enormous doses of cells were injected repeatedly into the fathers from birth until the end of the breeding period. Heroically, Brent and his co-workers repeated this anomalous procedure, but they made one mistake. The very first injection of cells, given on the day of birth, consisted of a total of 10^8 composed of equal quantities of bone marrow and spleen cells in Gorczynski and Steele's experiments, and 5×10^7 composed of one part of marrow to ten parts of spleen in Brent and colleagues' experiments. Brent's group is inclined to minimize the significance of this difference, but this seems a foolhardy attitude. Since there was, until Gorczynski and Steele's results were published, no reasonable expectation that any protocol would work, Brent and colleagues cannot dismiss any substantial feature of the successful protocol as insignificant. Curiosity and scepticism are, I am glad to read, carrying these authors still further to the point of reproducing even this aspect of the original protocol in their future experiments.

The importance of this difference in protocol is impossible to assess at present, but there is one crucial point which deserves more emphasis than Brent's group give to it. Amazingly, despite the ruthless régime of tolerance induction, the experimental fathers were apparently not fully tolerant by the measure of their ability to produce *in vitro* cytotoxic activity against the tolerizing parent. In Gorczynski and Steele's experiments the tolerant fathers were shown to be as tolerant as normal F_1

hybrids. Gorczynski and Steele's *in vitro* induction system was less efficient than that of Brent and colleagues, and weak residual activity may not have been visible, but the difference between apparently fully tolerant and apparently not fully tolerant fathers cannot be considered trivial in an experiment on the transmission of tolerance. Perhaps the difference is real and important, and perhaps it too hinges on the difference in the cellular composition of the first inoculum used to induce tolerance.

When Gorczynski and Steele's first results were published, discussion among colleagues threw up two further kinds of experiment which we felt would explore the generality of the original claim. As self-tolerance is presumably an acquired characteristic, the A homozygous progeny of an (A × B)F_1 father mated to an A mother might show inherited hyporesponsiveness to B alloantigens. Second, since the method of inducing tolerance used by Gorczynski and Steele was so unusual and must have resulted in a high level of chimaerism in the tolerant fathers, we wondered idly what would be the immunological behaviour of the progeny of male A⟷B embryo fusion chimaeras with extensive chimaerism in all tissues, ideally including the gonads as well. So rapid and relentless is the march of science that both our lunchtime speculations have already appeared as substantial experimental results. Hasek and colleagues[8] have done the backcross experiment and McLaren and colleagues (including some of Brent's group) the embryo fusion chimaera experiment[9]. In neither case was any detectable level of specific hyporesponsiveness found in the progeny of the appropriate matings. Again, therefore, Steele's hypothetical mechanism finds itself narrowly constrained by results. If acquired immunological tolerance can be inherited through the male, then the conditions under which this occurs are special and not yet adequately defined. Weissmann and Mendel still provide the general case. We need an explanation for Gorczynski and Steele's tropical Volute shell, but the superficial deposits of the Midland counties are still of glacial origin.

There is much good in this fascinating transaction. A novel and imaginative hypothesis has provoked a series of experiments which would probably never otherwise have been done, certainly not with so much care. The results are conflicting, but there is a *prima facie* case from Gorczynski and Steele's two successful studies for continuing to treat the matter as being of the greatest importance. Even if it turns out, surprisingly, that both sets of positive results have some completely trivial explanation, our perspective will still have been enduringly enriched by the reminder that sex may not be the only way of transferring DNA between cells.

There is also something bad about what has happened over the past few months, which concerns the fragile and ill articulated matters of tact and style in the presentation of scientific material. Steele's book, *Somatic Selection and Adaptive Evolution*, was unlikely to have turned many heads without some truly original experimental support. It conflates an unacceptably wide variety of experimental findings into a single scheme, and fails in its obligation to apply an intensive critical appraisal in conventional terms to each set of findings in turn. The issue with which it deals, the mechanism of adaptive evolution, is not an entirely open question. Recent evidence for great complexity in the organization of genetic material shows that the beguiling simplicities of the double helix and the genetic code are probably not sufficient to explain all the properties of evolving systems; but this does not mean that all that has gone before can be discarded. Natural selection remains in the pre-eminent position that Darwin and Wallace created for it—as a mechanism for adaptive evolution. One of the most important subsidiary conclusions of genetical research in the last half-century is the recognition that developmental plasticity or individual adaptability to environmental heterogeneity is an evolved character under genetic control. C. H. Waddington[10] pointed out as long ago as 1942 that genetic modifications to adaptable systems can simulate the inheritance of acquired characters. In relegating discussion of this fundamentally important concept to a dismissive footnote, Steele does both himself and modern genetics a disservice. It is the current fashion to level popular attacks at the theory of evolution. Twice in the past year Darwin could be found staring gloomily from one corner of the *Sunday Times Magazine*, whilst first Lamarck and then Professor Stephen Gould of Harvard University confronted him from the opposite corner. Whether Professor Gould would wish to represent the opposition to classical Darwinism seems doubtful, but the earlier confrontation with Lamarck was designed to represent the challenge raised by Steele's work. If Steele is feeling the heat of an intensive critical scrutiny, both of his ideas and of his experiments, it is because, like Icarus, he has flown close to the Sun.

ADDENDUM

Since this article was written there have been at least three more published attempts[11-13] to reproduce Gorczynski and Steele's findings. One of these[13] is the work promised by Brent and colleagues in which the conditions for the induction of tolerance used by Gorczynski and Steele in their published description are accurately followed. Furthermore, the state of tolerance of the experimental fathers has now been found to be the same as that of a normal $(A \times B)F_1$. None of these new attempts to reproduce paternal transmission of tolerance has succeeded. The reasonable judge would now surely admit that the issue has been treated with the seriousness due to it, and would probably conclude that Gorczynski and Steele were 'wrong' in some unspecified way. My own view is that the only investigators who could justify continuing to work on the issue out of public funds are Gorczynski and Steele themselves. The onus is clearly on them to explain why so many distinguished laboratories are unable to repeat their work. This will take either a clear demonstration of the critical variable, or a frank admission that, as many people now fear, the original results were faulty.

1. Darwin, C. *Life and Letters* Vol. 1 (ed. Darwin, F.) 56 (John Murray, London; 1888).
2. Steele, E. J. *Somatic Selection and Adaptive Evolution* (Williams-Wallace International and Croom Helm, Toronto and London; 1979).
3. Gorczynski, R. M. & Steele, E. J. *Proc. natn. Acad. Sci. U.S.A.* **77**, 2871–2875 (1980).
4. Gorczynski, R. M. & Steele, E. J. *Nature* **289**, 678–681 (1981).
5. Darwin, C. *More Letters* (eds Darwin, F. & Seward, A. C.) 176 (John Murray, London; 1903).
6. Darwin, C. *Variations in Animals and Plants under Domestication* Vol. 2 (John Murray, London; 1868).
7. Brent, L., Rayfield, L. S., Chandler, P., Fierz, W., Medawar, P. B. & Simpson, E. *Nature* **290**, 508–512 (1981).
8. Hasek, M., Holan, V. & Chutna, J. in *Cellular and Molecular Mechanisms of Immunological Tolerance* (eds Hraba, T. & Hasek, M.) (Marcel Dekker, New York, in the press).
9. McLaren, A., Chandler, P., Buehr, M., Fierz, W. & Simpson, E. *Nature* **290**, 513–514 (1981).
10. Waddington, C. H. *Nature* **150**, 563–565 (1942).
11. Smith, R. N. *Nature* **292**, 767–768 (1981).
12. Nisbet-Brown, E. & Wegmann, T. G. *Proc. natn. Acad. Sci. U.S.A.* **78**, 5826–5828 (1981).
13. Brent, L., Chandler, P., Fierz, W., Medawar, P. B., Rayfield, L. S. & Simpson, E. *Nature* **295**, 242–244 (1982).

Jonathan C. Howard is at the ARC Institute of Animal Physiology, Cambridge, UK.

This article first appeared in *Nature* Vol. **290**, pp. 441–442; 1981.

CHAPTER FOUR

THE PATTERN OF NATURE

A strange feature of Darwin's life is that, after conceiving the idea of evolution by natural selection, he did not at once rush into print, but waited for twenty years, and then only published because his hand was forced when Wallace conceived the same idea. He did many things in the meanwhile, but probably the most time-consuming was his description and classification of barnacles. This was not an irrelevant exercise. If his theory of evolution was correct, then it should be possible to interpret existing patterns of variation as being the result of an evolutionary process. In studying barnacles, he was finding out whether this is so.

The oddest controversy which has occurred in evolutionary biology in recent years has broken out among taxonomists, whose business it is to classify organisms. The architects of the 'modern synthesis' in the 1940s were much concerned with the 'species problem'. A major contribution of men like Mayr, Rensch and Stebbins was to show that, although sexually reproducing organisms in a given region do fall into a series of discrete 'species', between which intermediates are rare or absent, this apparent uniqueness largely breaks down when one studies plants or animals over a wide geographic area. To quote Mayr, 'In every actively evolving genus we find that there are populations that are hardly different from each other, others that are as different as subspecies, others that have almost reached species level, and finally still others that are full species'.

This question, of the nature of species, is of obvious relevance to the debate about 'punctuation' (pp. 125–181), but the 'cladistics' debate is mainly concerned with another matter: how are species to be classified into higher taxa—genera, families, orders, classes and so on? The term 'cladistics' takes its origin in the work of Hennig, who, as a convinced evolutionist, argued that classification should reflect 'phylogeny'—that is, the evolutionary relationships of organisms. This faced him with two problems. First, how can one best deduce phylogenetic relationships? Second, given that one has reached a hypothesis about the phylogeny of a group of organisms, how should one classify them? The first of these questions concerns the external world; the second is purely semantic—it concerns the names one gives to things.

Hennig made two interesting suggestions about how one should deduce phylogenies. The first is that for each species, X, in a set of species, one should attempt to find a 'sister group' Y, such that X and Y have a common ancestor more recently in the past than either do with any other member of the set. The second is that, in identifying sister groups, one should pay attention, not equally to all similarities, but specifically to similarities in derived traits ('apomorphisms') rather than in primitive traits ('plesiomorphisms'). Thus men, bats and lizards have five fingers, but this is not evidence of close relationship, because the condition is primitive in land vertebrates. Horses and zebras have one finger; this *is* evidence of close relationship, because to have one finger is not primitive. The argument is convincing; its critics point out, however, that it is not always so easy to decide which state is primitive and which derived.

It is my impression—and I am only an outsider in this debate—that, despite the difficulties, these ideas of Hennig's were a genuine contribution to the elucidation of phylogenies. The main criticisms of Hennig from the architects of the modern synthesis (particularly Mayr and Simpson) concerned his answer to the second question—how to convert a phylogeny into a classification. He insisted that a 'taxon' (for example, a family or a class) should include all the descendents of a single ancestor. This would rule out the class 'Reptilia' as commonly understood (that is, as a class including lizards, crocodiles and tortoises, but not birds or mammals), because the common ancestor of the reptiles was also an ancestor of the birds and mammals, which therefore, by Hennig's principles, should be included in the same taxon.

It will be seen that this is an argument about names, not about the world. As such, I refuse to get excited about it, although I can see some practical force to the objections raised by Simpson. If this were the only issue, the following articles by Patterson and Charig would not appear in this book. The reason for the present excitement at the British and at the American Museums of Natural History is that a few of Hennig's followers have become 'transformed cladists'. They have discovered that Hennig's procedures for constructing 'cladograms' (trees representing relationships) can be carried out quite satisfactorily without thinking about evolution. From this, they conclude that maybe evolution is not a necessary hypothesis in biology, and that evolutionary arguments are 'mere rhetoric'.

Now it is obviously true that one can apply Hennig's methods without thinking about evolution. It is also true that one can drive a motor car without understanding how internal combustion engines work. However, if internal combustion engines didn't work, one would not be able to drive cars, and if evolution hadn't happened one would not be able to apply Hennig's methods successfully. It is not merely that the concepts of sister group, and of primitive and derived traits, make little sense in the absence of an evolutionary hypothesis. The crucial point is that Patterson's first 'axiom' of cladistics—that 'features shared by organisms manifest a hierarchical pattern in nature'—is true only because organisms evolved. Thus it is also true of languages, which evolved by a branching process. It is not true of the chemical elements, which did not evolve, and which, as Mendeleyev showed, manifest a periodic pattern in nature. A contemporary of Darwin's, McCleay, believed that organisms were best classified in a set of interlocking pentagons. The idea seems ridiculous today only because we are evolutionists.

The transformed cladists, then, are in the illogical position of saying that evolution is irrelevant to taxonomy, while accepting as an axiom a proposition which is true only because evolution happened. Darwin knew better. The diagram of a tree of nature appeared in his notebooks as early as 1838, and it is the only diagram in the *Origin*.

Cladistics

COLIN PATTERSON

CLADISTICS is a method of systematics. Other names for the method are phylogenetic systematics (from the title of Willi Hennig's 1966 book[1]), Hennigian systematics, and cladism. 'Clade' is a term introduced by Julian Huxley in 1957 for 'delimitable monophyletic units', and at its simplest cladistics is a technique for characterizing (delimiting) a hierarchy of groups. Of course, the same is true of Linnean systematics, or of numerical taxonomy. Yet whenever and wherever cladistics is discussed, there is controversy. The virulence of this controversy and the partisan feelings evoked are remarkable in so old and apparently harmless a discipline as systematics; there has hardly been such bad temper in biology since the violent arguments provoked 120 years ago by Darwin's theory of evolution by natural selection. The aims of this article are to present a personal view of the principles and applications of cladistics, and to discuss the reasons for the controversy it has provoked.

THE METHOD

The axioms of cladistics are only three:
1. Features shared by organisms (homologies) manifest a hierarchical pattern in nature.
2. This hierarchical pattern is economically expressed in branching diagrams, or cladograms.
3. The nodes in cladograms symbolize the homologies shared by the organisms grouped by the node, so that a cladogram is synonymous with a classification (figure 1).

Consider table 1 which lists the amino acids present at various positions in the myoglobin chain of four different species. Of these 20 characters, 1–4 are the same in all four species; 66 is different in all four; the remainder are the same in two or three of the four. With four taxa (A, B, C, D) there are 12 possible sorts of shared characters:

Shared by four: ABCD – 1, 2, 3, 4

Shared by three: ABC – 5, 9, 26, 30, 34
                 ABD – 59
                 ACD – 27
                 BCD – none

Shared by two: AB – 12, 35, 48
                AC – 13, 22
                AD – 19, 21
                BC – 74
                BD – 13
                CD – 48

Shared by none: A, B, C, D, – 66

(For three taxa there are five possible types of character; for five there are 27, and so on).

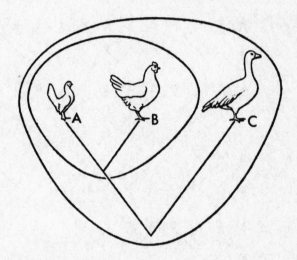

Fig. 1. *A cladogram and its meaning. The two chickens are linked by a node expressing shared homologies absent in the goose. The chickens form a group AB to which the goose does not belong. The chickens and the goose share homologies characterizing a more extensive group, ABC, of which AB and C are subgroups. After Zimmermann[2].*

Characters

Taxa	1	2	3	4	5	9	12	13	19	21	22	26	27	30	34	35	48	59	66	74
A	G	L	S	D	G	L	N	V	A	I	P	Q	E	I	K	G	H	E	A	G
B	G	L	S	D	G	L	N	I	T	V	G	Q	D	I	K	G	H	E	I	N
C	G	L	S	D	G	L	K	V	G	L	P	Q	E	I	K	T	G	A	G	N
D	G	L	S	D	Q	Q	T	I	A	I	A	H	E	M	H	D	G	E	Q	A

Table 1. *Twenty characters of the four species A, B, C and D. See text for explanation.*

Figure 2 sets out the 15 possible combinations of the 12 types of character in table 1. All 15 cladograms agree with the ABCD characters (1–4) and the A, B, C, D character (66), for characters of these types say only that there are four taxa and that they form a group. The cladograms differ in the success with which they present the types of character uniting two or three of the taxa. The most successful is cladogram 2.11, which includes eight of the 17 characters uniting two or three taxa. Cladogram 2.12 includes seven of these characters, 2.10 includes six, and the remainder from one to four. By this criterion, success in summarizing the data of table 1, we should prefer cladogram 2.11, with 2.10 a close runner-up.

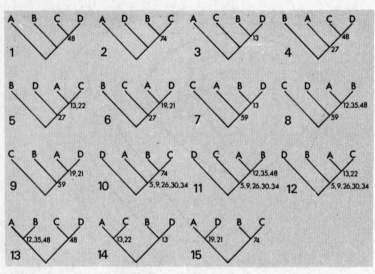

FIG. 2. *The 15 possible dichotomous cladograms for four taxa, and the shared characters in table 1 expressed by each.*

To evaluate that choice, we need more information. In fact, the taxa in table 1 are *Homo sapiens* (A), the red kangaroo, *Megaleia rufa* (B), the platypus, *Ornithorhynchus anatinus* (C), and the chicken, *Gallus gallus* (D). Given this information the biologist can bring experience to bear. He 'knows' that cladogram 2.11 is the correct one, for it was taught at school, and can supply further characters corroborating it, such as:

ABC – hair, mammary glands, ear ossicles
AB – vivipary, teeth, spiral cochlea, teats.

Of course, one can also supply characters of other types, such as:

ABD – bipedal locomotion
AC – five digits in hind limb
BC – tail
BD – four digits in hind limb
CD – horny bill, ovipary, cloaca.

But the biologist 'knows' these are not true characters bearing on the interrelationships or grouping of the four animals. How was that knowledge acquired?

The 'false' characters just listed fall into two categories. Some, like the bill of birds and platypus, and the bipedality of bird, kangaroo, and human, are rated as false because they are 'not really the same'—they are non-homologous, chance or superficial similarities, or in evolutionary terms, convergent. The basis for this conclusion is that the groups specified by such characters (birds, platypus, and turtles as beaked animals; birds, humans, and kangaroos as bipedal animals) are contradicted by many other characters and corroborated by none. In other words, these characters fail the principal test of homology, congruence with other characters. Some of the shared characters in table 1, such as the alanine shared by human and chicken at position 19, can be rated as non-homologous, or 'not really the same' by observation of the distribution of the amino acid in other species.

The second class of 'false' characters includes such things as five toes in man and platypus, the tail in kangaroo and platypus, or ovipary and cloaca in bird and platypus. We regard these as homologies, as 'really the same'. But these homologies are irrelevant to the grouping of the four species. The reason is that the groups specified by these homologies include other animals besides birds and mammals: ovipary characterizes Metazoa, a postanal tail Chordata, the cloaca Craniata, and five toes characterize Tetrapoda. In terms of the four species in table 1, these are all ABCD characters, what Hennig called symplesiomorphies (shared primitive characters). The generality of such characters, and the problems of grouping to which they are relevant, are more inclusive than the problem at hand. Yet there is an obvious contradiction here. For instance, five toes are listed above as a man/platypus character, but I have just asserted that it is really an ABCD character. The contradiction is resolved through ontogeny, for the embryo bird and kangaroo have rudiments of five toes, and the four-toed condition arises by ontogenetic transformation. We recognize a general condition (five toes, long tail, cloaca) and a special condition (four toes, rudimentary tail, separate anus and urogenital opening) of more restricted distribution. Animals with the special condition also have the general condition, in early ontogeny.

Ovipary cannot be explained away like this, for the embryonic human and kangaroo are not oviparous. The basis for calling ovipary 'really' an ABCD character is the same as that for regarding the bill of bird and platypus as 'not really' a CD character: correlation with other homologies. For treating ovipary as an independent homology—grouping all oviparous animals—is contradicted by the host of characters showing that oviparous mammals (monotremes) should be grouped with viviparous mammals, and that oviparous sharks and insects should each be grouped with viviparous sharks and insects. The contradiction is resolved by character evaluation: ovipary is a general condition (eggs are produced) and vivipary is a special condition (eggs hatch within the body) defining subgroups within the group of oviparous animals.

Referring back to table 1, several of the shared characters can be treated as irrelevant (symplesiomorphous) by observing their distribution in other organisms. For instance, at position 13 kangaroo and chicken share isoleucine, but the same amino acid occurs at this position in all the reptiles in which myoglobin has been sequenced. It is therefore a general condition, irrelevant to groupings within birds and mammals—an ABCD character. The contradiction between that statement and the BD distribution of the character in table 1 cannot be resolved by ontogeny (like the four toes of chicken and kangaroo), for in man and platypus there is no ontogenetic transformation of isoleucine into valine. Nor can the contradiction be resolved by reevaluation of the characters (as with ovipary), for valine cannot be regarded as a special condition of isoleucine. The only explanation I can imagine is phylogenetic, that during the history of man and platypus an isoleucine codon (e.g. TAT) has been transformed into a valine codon (e.g. CAT).

Of course, it is possible that some of the myoglobin characters treated as true homologies in cladogram 2.11 (figure 2) could also be argued away as non-homologous or irrelevant. As an example, I included 'teeth' amongst the morphological man/kangaroo characters on p. 112. Considering only humans, kangaroos, platypus, and chicken, teeth are a man/kangaroo character, but of course, teeth characterize Gnathostomata: the character is a symplesiomorphy irrelevant amongst birds and mammals. The absence of teeth in birds and platypus is more interesting. Lack of teeth is often cited as a character of birds, of Recent monotremes and other gnathostome groups. But how can absence of something characterize a group?

Table 2 presents some more myoglobin data for four species: A is *Homo sapiens*, B is a reptile, alligator, C is a teleost, tuna, and D a shark, *Heterodontus*. As with the data in table 1, the different types of character can be listed. If absence is treated as a character, table 2 contains these six types:

ABCD – 10   AB – 2,3,4
ABC – 11   CD – 1,2,3,4
ACD – 0   A,B,C,D – 5

And if presence alone is counted, table 2 contains four types:

ABCD – 10
ABC – 11
AB – 2,3,4
A,B,C,D – 5

The cladograms that best summarize these two sets of characters are shown in figure 3. The first, including both presence and absence, contains four of the six types of character in the list: the ABC and ACD at positions 11 and 0 are incongruent with it. The second cladogram, based on presence, contains all the characters in the list, and is a complete summary of the data. The two cladograms are different. The first groups tuna with shark, the second groups tuna with man and alligator. Which is correct?

Here, for the first time, we come up against a controversial aspect of cladistics. For some biologists assert that the first cladogram is correct,

Characters

Taxa	0	1	2	3	4	5	10	11
A	-	G	L	S	D	G	V	L
B	M	E	L	S	D	Q	V	L
C	-	-	-	-	-	A	V	L
D	-	-	-	-	-	T	V	N

TABLE 2. *Amino acids at certain positions in the myoglobin chain of A,* Homo sapiens, *B, alligator, C, tuna, and D, the shark* Heterodontus.

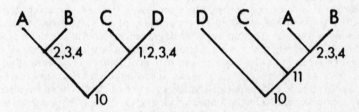

FIG. 3. *Cladograms summarizing the data of table 2. Left, if presences and absences are counted; right, counting presences alone.*

and others (including me) insist on the second. The grouping of tuna and shark, or in more general terms, of chondrichthyans and bony fishes (other sharks and bony fishes also lack amino acids at positions 1–4) corresponds to the class Pisces of Linnean and classical taxonomy, and some biologists continue to find this a useful concept. But on the basis of table 2, the only characters of this group are absences. A few minutes thought will show that this is generally true, for morphology as well as for the few amino acids in table 2: fishes, as a group, can only be defined by the absence of characters. Scales, fins, gills, and anything else one might think of as characteristic of fishes are not—those mentioned are all found in amphibians, for example.

Pisces, a group characterized only by lack of characters, is an example of what Hennig called a paraphyletic group, in his definition, one based on symplesiomorphies, or shared primitive features. Here, I have classed symplesiomorphies not as primitive, but as irrelevant characters. They are irrelevant because the groups specified by such characters are more extensive than the problem under study. Absences have the same status.

In figure 3 the shark and teleost are grouped by absence at positions 1–4 in myoglobin. But other organisms, such as *Escherichia coli* and the buttercup, also lack amino acids at these positions (they lack myoglobin altogether). The group of organisms lacking amino acids at positions 1–4 in myoglobin would include, so far as we know at present, the whole of life except amniotes, a grouping of no interest to anyone. In short, the only groups of interest in systematics are those which have characters; uncharacterizable (paraphyletic) groups are not groups.

There are two general reasons for preferring that cladogram which best summarizes a set of data (e.g. no. 11 in figure 2) as a general theory about the interrelationships or classification of the organisms concerned. The first is parsimony: cladogram 2.11 is more parsimonious than the others because it requires one to argue away, or neglect, fewer characters in table 1, and this is also true when the data are expanded to include all our knowledge of the morphology, physiology, etc. of birds, mammals, and other organisms. It is worth emphasizing that parsimony, as a criterion, has nothing to do with opinions about evolution (whether it follows the shortest course or not) or any other process in biology. We accept the most parsimonious, or simplest explanation of the data because parsimony is the only criterion available. If parsimony is set aside, there can be no other reason than personal whim or idiosyncrasy for preferring any one of a myriad possible explanations of a set of data. Figure 2 shows that for four taxa there are 15 possible dichotomous cladograms, each equivalent to a different explanation of a set of data. There are also trichotomous and tetrachotomous cladograms, making 26 possibilities for four taxa; for ten there are 282 137 824 (ref. 3). Without the parsimony criterion, could we ever make sense of nature?

The second reason for preferring cladogram 2.11 may be called heuristic or predictive. We can test the cladogram by its prediction that further data, not yet studied, will conform to this pattern. This cladogram, in the general form that marsupials will share characters with placentals, rather than with monotremes and birds, and that monotremes will share characters with therians rather than with birds, has been tested and corroborated repeatedly since 1834, when de Blainville first grouped mammals in this way. The sequences of myoglobin and other proteins are only the most recent, and perhaps the most sensitive, of these tests.

The cladistic method forces systematists to be explicit about the groups they recognize, and the characters of those groups—that is, the characters they regard as homologies. The main advantage of this is that systematics need no longer rely on authority, tradition, and idiosyncrasy, irrational criteria which are notorious in the discipline. Explicit systematics in the form of dichotomous cladograms (dichotomies preferred because of their greater information content and testability[4]) can readily be understood, criticised, and tested, and this seems to be the only way in which systematics can be brought within the general framework of scientific method.

On a more general theme, cladistics is not limited to biological systematics. It deals with homologies, their evaluation and parsimonious interpretation, and the method is applicable in any field in which relations comparable to homology can be recognized. The method is already applied in linguistics and textual criticism[5], where the homologies are features shared by tongues or texts. It is beginning to be applied in biogeography[6], another discipline notorious for idiosyncracy, or lack of an agreed method. Here the homologies are sister-taxa shared by areas. And it may be applicable in geology (figure 4) and other historical sciences.

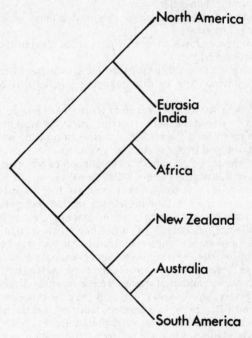

FIG. 4. *A geological area cladogram, summarizing current opinion on the sequence of break-up of Pangea. After Nelson and Platnick[6].*

THE CONTROVERSY

I turn now to the second part of this article, the reasons for the controversy that has surrounded cladistics since its widespread introduction in the late 1960s (see ref. 7, for a history). I hope the reader will have found my summary of the method uncontroversial, even self-evident or commonplace. Yet what might be called the biological establishment has been almost unanimously negative in its response[8-12]. Meanwhile, a growing band of enthusiasts has refined the method[4, 13-18], and for 15 years the argument has continued, chiefly in the journal *Systematic Zoology*, but surfacing elsewhere with increasing frequency. Mayr's 1974 paper[10] is the most fully argued criticism of cladistics. The paper, and the replies to it[19-21] should be read by those interested in following up the debate, but Mayr's main points may be summarized as follows:

1. Cladists have altered the meaning of terms like phylogeny, relationship, and monophyly, by defining them in terms of common ancestry.
2. Cladists neglect anagenetic change (rates of evolution) by concentrating on splitting of lineages.
3. Cladists adopt a purely formalistic species definition, and insist that speciation is dichotomous.
4. The ranking procedures advocated in cladistic classification are unsound and impractical.
5. Cladists neglect the difficulty of deciding whether characters are primitive or derived, and of discriminating parallel and convergent evolution.

These criticisms may seem strange to those unfamiliar with the controversy who have read the first part of this article, for most of them refer to topics not mentioned there. This is a reflection of the way cladistic theory has developed over the last few years, and the direction of that development should be plain from a comparison of Mayr's points with the first part of this article. Hennig's 1966 book, as the title *Phylogenetic Systematics* suggests, was based in evolutionary theory, just as Mayr's criticisms are almost all to do with aspects of that theory (speciation, anagenetic change, parallelism, common ancestry, etc.). But as the theory of cladistics has developed, it has been realized that more and more of the evolutionary framework is inessential, and may be dropped. The chief symptom of this change is the significance attached to nodes in cladograms. In Hennig's book, as in all early work in cladistics, the nodes are taken to represent ancestral species. This assumption has been found to be unnecessary, even misleading, and may be dropped. Platnick[4] refers to the new theory as 'transformed cladistics' and the transformation is away from dependence on evolutionary theory. Indeed, Gareth Nelson, who is chiefly responsible for the transformation, put it like this in a letter to me this summer: 'In a way, I think we are merely rediscovering preevolutionary systematics; or if not rediscovering it, fleshing it out.'

Mayr's and Simpson's criticisms (see quotes in Platnick[4], for the latter) assume that cladistics is to do with evolution. But cladistics, as I have tried to show, is not necessarily about evolution—speciation, ancestry, and such things. It is about a simpler and more basic matter, the pattern in nature—groups, hierarchies or nested sets of groups, and characters of groups. Groups which are uncharacterizable, or are characterizable only by lack of characters, are not groups. Yet evolutionary systematics, as described, for instance, in Simpson's 1961 book[22] and Mayr's 1974 paper[10], makes no distinction between the two kinds of group. In particular, evolutionary systematists are interested in 'ancestral' groups, and the chief glories of more than a century of evolutionary systematics are held to be the unravelling of phylogenies, particularly within the

vertebrates, where the conclusions are encapsulated in general statements such as 'reptiles were ancestral to birds and mammals', or more
specific statements such as 'tetrapods evolved from rhipidistian fishes in
the Devonian'. Such statements are meant to convey information, yet to
the cladist (and presumably to anyone else who cares to analyse them)
the only information conveyed is that the groups held to be descendent
(vertebrates, tetrapods, birds, mammals, for instance) are characterizable,
and the groups held to be ancestral are not, for all such groups (invertebrates, fishes, reptiles, rhipidistians, therapsids, for example) are, as
shown above for fishes, characterized only by lack of things.

Why should evolutionists wish to retain these non-groups (fishes,
reptiles, etc.) when they agree with cladists that the characters of the
organisms contained in them dictate different groupings: for example,
that lungfishes and coelacanths go with tetrapods; that some reptiles go
with birds and some with mammals? The answer given by evolutionists
is that these groups express adaptation—they are[12] 'adaptively unified'
or[10] 'well characterized grades'. 'Grade' is another term popularized by
Huxley in 1957, for 'delimitable and persistent anagenetic units'; as
Mayr[10] put it 'all members of a grade are characterized by a well integrated
adaptive complex'. These groups then, like extinct groups such as
rhipidistians and therapsids, are meant to express something about the
process of evolution, rather than the pattern of character distribution. But
how can the process of evolution be discerned except through the pattern
of character distribution? As shown in the first part of this article, analysis
of character distribution leads to recognition of a hierarchy of groups,
which may be represented by a cladogram or classification expressing the
most parsimonious, or rational explanation of that distribution. That
explanation involves reassessment of some characters either as 'not really
the same' (non-homologous; convergent or parallel in evolutionary
terms) or as irrelevant (homologous but irrelevant to the problem at
hand; primitive in evolutionary terms), and observed character distribution is the reason for those reassessments. How can evolutionary theory
require a different interpretation of the pattern of character distribution?

In my view, the most important outcome of cladistics is that a simple,
even naive method of discovering the groups of systematics—what used
to be called the natural system—has led some of us to realize that much
of today's explanation of nature, in terms of neo-Darwinism, or the
synthetic theory, may be empty rhetoric. Eldredge and Cracraft[23] provide
a detailed criticism of the transformational approach in biology, and the
ensuing interest in story-telling about adaptive change. They argue that
cladistic theories of pattern are the only available tests of macroevolutionary theory. In Platnick's[4] words 'what Hennig may well have done
. . . is to demonstrate the inadequacy of the syntheticist paradigm, by
showing that we are hardly likely to achieve any understanding of the
evolutionary process until we have achieved an understanding of the
pattern produced by that process, and that even today we have hardly
begun to understand the pattern'. If this is the result of cladistics, to give
neo-Darwinism a good shake, then perhaps its critics are right to get
excited.

1. Hennig, W. *Phylogenetic Systematics*. Urbana: University of Illinois Press (1966) (Second edition, 1979).

2. Zimmerman, W. *Evolution. Die Geschichte ihrer Probleme und Erkenntnisse*. Freiburg & München: Karl Alber (1953).

3. Felsenstein, J. The number of evolutionary trees. *Systematic Zoology*, **27**, 27–33 (1978).

4. Platnick, N. I. Philosophy and the transformation of cladistics. *Systematic Zoology*, **28**, 537–546 (1980).

5. Platnick, N. I. & Cameron, H. D. Cladistic methods in textual, linguistic, and phylogenetic analysis. *Systematic Zoology*, **26**, 380–385 (1977).

6. Nelson, G. J. & Platnick, N. I. A vicariance approach to historical biogeography. *Bioscience*, **30**, 339–343 (1980).

7. Dupuis, C. La 'Systématique Phylogénétique' de W. Hennig (Historique, discussion, choix de références). *Cahiers des Naturalistes*, **34**, 1–69 (1979).

8. Cain, A. J. One phylogenetic system. *Nature*, **216**, 412–413 (1967).

9. Darlington, P. J. A practical criticism of Hennig-Brundin 'phylogenetic systematics' and Antarctic biogeography. *Systematic Zoology*, **19**, 1–18 (1970).

10. Mayr, E. Cladistic analysis or cladistic classification? *Zeitschrift für zoologische Systematik und Evolutionsforschung*, **12**, 94–128 (1974).

11. Simpson, G. G. Recent advances in methods of phylogenetic inference. In *Phylogeny of the Primates*, ed. Luckett, W. P. & Szalay, F. pp. 3–19. New York: Plenum (1975).

12. Van Valen, L. Why not to be a cladist. *Evolutionary Theory*, **3**, 285–299 (1978).

13. Brundin, L. Transantarctic relationships and their significance. *Kungliga Svenska Vetenskapsakademiens Handlingar*, (4) **11**, 1, 1–472 (1966).

14. Nelson, G. J. Classification as an expression of phylogenetic relationships. *Systematic Zoology*, **22**, 344–359 (1973).

15. Nelson, G. J. Ontogeny, phylogeny, paleontology and the biogenetic law. *Systematic Zoology*, **27**, 324–345 (1978).

16. Nelson, G. J. Cladistic analysis and synthesis: principles and definitions, with a historical note on Adanson's 'Familles des Plantes' (1763–1764). *Systematic Zoology*, **28**, 1–21 (1979).

17. Griffiths, G. C. D. On the foundations of biological systematics. *Acta Biotheoretica*, **23**, 85–131 (1974).

18. Bonde, N. Cladistic classification as applied to vertebrates. In *Major Patterns in Vertebrate Evolution*, ed. Hecht, M. K., Goody, P. C., & Hecht, B. M. pp. 741–804. New York: Plenum (1977).

19. Nelson, G. J. Darwin-Hennig classification: a reply to Ernst Mayr. *Systematic Zoology*, **23**, 452–458 (1974).

20. Rosen, D. E. Cladism or gradism? A reply to Ernst Mayr. *Systematic Zoology*, **23**, 446–451 (1974).

21. Hennig, W. 'Cladistic analysis or cladistic classification': a reply to Ernst Mayr. *Systematic Zoology*, **24**, 244–256 (1975).

22. Simpson, G. G. *Principles of Animal Taxonomy*. New York: Columbia University Press (1961).

23. Eldredge, N. & Cracraft, J. *Phylogenetic Patterns and the Evolutionary Process*. New York: Columbia University Press (1980).

Colin Patterson is at the British Museum (Natural History).

This article first appeared in *Biologist* Vol. **27**, pp. 234–240; 1980. It is reproduced with permission.

Cladistics: a different point of view

ALAN CHARIG

COLIN Patterson's personal view of 'Cladistics' is a model of clarity in its presentation. As for its content, much of that is admirable too and factually unexceptionable; nevertheless I find my colleague's approach to systematics very different from my own and certainly not typical of the hundreds of systematists working in the British Museum (Natural History).

Essentially he is contrasting 'evolutionary systematics' with 'cladistics', at the same time pointing out that the cladistics of 1980 has developed a long way from the system put forward by Hennig in 1966—chiefly in that 'it has been realized that more and more of the evolutionary framework is inessential, and may be dropped.' He therefore believes that the main criticisms of cladistics (by people like Cain, Mayr, and Simpson) are now totally irrelevant, being directed against the antiquated views of Hennig on various topics connected with evolution and not against the 'non-evolutionary' cladistics of today.

I suggest that the true situation is represented much better if we recognize that the modern type of cladistics advocated by Patterson, what Platnick[1] calls 'transformed cladistics', is altogether different from Hennigian systematics and should be treated as a separate entity. In Hennigian systematics phylogeny is all-important and is conventionally depicted by a branching dendrogram (hence 'cladistics'—Greek 'klados', branch or young shoot). In 'transformed cladistics', however, evolution is deliberately ignored, being considered unproven and possibly unprovable; the customary use of dendrograms to represent the 'natural order' is therefore highly misleading in that it suggests progression with occasional dichotomies. 'Transformed cladistics' is neither Hennigian, phylogenetic, nor cladistic, and would be referred to more appropriately as 'natural order systematics'.

Natural order systematics demands an hierarchical arrangement of shared characters and of the organisms which possess them, preferred over all other possible arrangements on the sole criterion of maximal congruence or parsimony; such phenomena as parallelism and convergence or transformation series in stratigraphical sequences are rejected as being 'outside the scope of systematics'. The resulting arrangement is the 'natural order'—hence 'natural order systematics'—and is *ipso facto* the classification. The difficulty for those who are not advocates of this type of systematics, for people like me, lies in comprehending what the 'natural order' can possibly mean if it does not indicate any type of phylogenetic relationship. Certainly it indicates maximal congruence, for that is how it was arrived at; it consequently indicates the greatest aggregate ('overall') similarity, thus resembling a phenetic classification; and, to a creationist, it could be said to indicate the order in the mind of the Creator. But I cannot imagine what other 'relationship' can be shown by the shared possession of such a random assortment of characters—some due to common ancestry, some being similar adaptations to similar external causes, and others being purely fortuitous—unless we define 'related' as meaning nothing more than 'sharing common characters'! Not only is it impossible to disentangle those different types of character but, according to the dogma of natural order systematics, it is expressly forbidden to do so.

The true antithesis between the different types of systematics lies between what I call 'Simpsonian' (conventional evolutionary) systematics *plus* Hennigian systematics on the one hand and natural order systematics on the other. Simpsonian and Hennigian systematics are alike in that each comprises two distinct procedures. The first is an attempt to ascertain the phylogeny, an objective reality, and represent it by means of a branching diagram; the second is the production of an hierarchical classification based on that diagram. These two types of systematics might both be described as 'clado-evolutionary'. In both cases the attempt to discover the phylogeny is made by means of a character distribution analysis ('cladistic analysis'); preference is generally given to the most parsimonious arrangement, but other factors of an evolutionary nature are also taken into consideration, for example, the fossil record, parallelism and convergence. The Hennigian procedure involves a more formal, disciplined approach and is to be welcomed on that account.

Far greater differences exist in the derivation of a classification from the chosen arrangement. The Hennigian procedure is simply to base an hierarchical classification directly on the phylogenetic cladogram; no paraphyletic taxa are permitted, that is, each taxon must include all its own descendants, so that familiar taxa like Reptilia (which does not include Aves and Mammalia, descended from it) are unacceptable. The conventional Simpsonian method, on the other hand, is to divide up the phylogenetic 'tree' in arbitrary fashion so as best to reflect the present characters of the organisms as well as their genealogy, using 'grade' taxa as well as 'clade'. This produces a classification which is consistent with the phylogeny and in which each taxon is a single continuous segment of the phylogenetic 'tree', but which is by no means an exact reflection of the dendrogram.

It is easy to demonstrate the fallacy in Patterson's contention that paraphyletic taxa like Reptilia are 'non-groups' and that, in consequence, statements involving them—such as 'reptiles were ancestral to birds and mammals'—convey no useful information. A natural order or Hennigian systematist merely regards Aves and Mammalia as subordinate clades nested within the larger clade Amniota. Given an evolutionary interpretation, however, such a pattern implies that birds and mammals arose *within* the Amniota, presumably from Amniota which were not themselves birds or mammals—that is, by definition, from the reptiles! It must be admitted that a paraphyletic taxon is not equivalent to a clade, a natural monophyletic group, it is an artificial concept of the human mind; but it may still be defined in terms of a clade from which one or more younger, smaller clades nested within have been excluded. After all, a set which is defined by the presence of certain characters and by the absence of others is still a perfectly good set. And to recognize a paraphyletic taxon is a very convenient way in which to categorize a 'stem-group' or 'ancestral group'.

Incidentally, Patterson's so-called three 'axioms of cladistics' are not axioms at all. The first, 'Features shared by organisms (homologies) manifest a hierarchical pattern in nature' should really read 'hierarchical patterns', for, as the author himself makes clear, there is a choice of patterns—to be resolved by parsimony; and, thus modified, the statement is so general as to be meaningless. The other two 'axioms' are merely statements of the conventions used in graphical representation of the resulting pattern, which is synonymous with the classification.

Other aspects of 'cladistics' which could usefully be discussed include: the meanings of the words 'monophyly', 'polyphyly', etc.; the recognition of ancestors; recency of common ancestry; the delimitation of taxa with respect to time; and the absolute ranking of taxa. A very much longer and more detailed article covering all those points and many others is to be published shortly[2]. Meanwhile I might summarize my own attitude as follows:

1. I recognize that we have no absolute proof of the theory of evolution, by direct evidence of the senses; all the available evidence is merely circumstantial. However, there is no scientifically acceptable evidence against it and no other theory fits the known facts so well. I therefore accept it as a working hypothesis of immeasurable heuristic value.

2. The branching pattern of the phylogeny should be reconstructed as accurately as possible by means of a character distribution analysis conducted in a strictly disciplined Hennigian manner, modified and augmented in the light of evidence obtained from any other valid source, e.g. palaeontology.

3. The reconstructed phylogeny may be conveniently expressed, as completely as possible, by means of a branching diagram (dendrogram). *The dendrogram itself* provides a 'topographical' reference system in biology.

4. The dendrogram, representing continuity in time and space, may be divided into taxa in an arbitrary fashion in order to fulfil, as well as possible, the general requirements of a biological classification. Thus the classification should impart, as far as is consistent with division in that manner, the most important characteristics of each taxon at the time of its existence as well as the broad outline of its evolutionary history. Each taxon should correspond to a single continuous segment of the dendrogram; it must be of monophyletic origin, but it need not include all its own descendants, i.e. it may be paraphyletic.

I shall end my comments by adding that the virulence of the controversy mentioned by Patterson comes almost entirely from the 'cladists', some of whom (be they of the classical Hennigian persuasion or of the 'transformed' variety) embrace their creed with a quasi-religious fervour. Surely the subject of biological classification, as much as any other branch of science, merits a less emotional, more rational approach.

1. Platnick, N. I. Philosophy and the transformation of cladistics. *Systematic Zoology*, **28**, 537–546 (1980).
2. Charig, A. Systematics in biology: a fundamental comparison of some major schools of thought. In *Problems of Phylogenetic Reconstruction*, pp. 363–440, ed. Joysey, K. A. & Friday, A. E. Systematics Association Special Volume. London: Academic Press (1982).

Alan Charig is at the British Museum (Natural History).

This article first appeared in *Biologist* Vol. **28**, pp. 19–20; 1981. It is reproduced with permission.

CHAPTER FIVE

EVOLUTION—SUDDEN OR GRADUAL?

IN a recent book, *Darwin on Man*, Gruber has argued that the concept 'gradual = natural; sudden = miraculous' was a feature of Darwin's thinking from the time he was a student, remaining unchanged during his transition from a belief in the biblical account of creation to an acceptance of evolution. He may first have acquired it from Archbishop Summer's argument (on which he took notes when at Cambridge) that Christ must have been a divine teacher because, had he been merely human, his teachings could not so suddenly have transformed the Roman world.

An attack on the 'gradualist' interpretation of evolution, recently launched by a group of palaeontologists and summarised below in Gould's article, may therefore imply an attack on Darwinism itself. Before considering the nature of this attack, it is worth asking what, if any, is the connection between gradualism and Darwinism. It is, I think, as follows. For Darwin the essential task facing any theory of evolution was to explain the detailed adaptation of organisms to their ways of life. If this adaptation is to be brought about by the natural selection of variants which are in their origin non-adaptive, the process must involve a very large number of steps, many of them small in extent. To produce a detailed adaptation by means of mutations of large effect only—macromutations—would face the same difficulties as would a surgeon obliged to perform an operation using a mechanically controlled scalpel which could only be moved a foot at a time. Fisher, in an argument quoted below by Lande, went further; he argued that, since existing organisms are well adapted, large mutations are necessarily harmful. I am not fully persuaded of this. I agree that adaptation could not be produced by the selection of macromutations only, but I cannot see why occasional macromutations should not have been incorporated by selection. The question is ultimately an empirical one; Lande mentions some of the empirical evidence that most changes have been 'gradual', in the sense of depending on many small changes.

Both for Darwin and for neo-Darwinists, then, it is central that the adaptation of organisms to their ways of life is, in the main, brought about by the natural selection of numerous genetic differences between the members of populations. However, it has never been part of the theory that evolution proceeds at a constant rate. Darwin himself wrote, in the *Origin*, 'Although each species must have passed through numerous transitional stages, it is probable that the periods during which each underwent modification, though many and long as measured by years, have been short in comparison with the periods during which each remained in an unchanged condition.'

The punctuationists have in fact made two claims. The minor claim is an empirical one: it is that the fossil record reveals long periods of little or no evolutionary change, punctuated by short periods of rapid change, usually associated with lineage splitting (that is, the division of a single ancestral species into two daughter species). Williamson's paper documents this process convincingly for the molluscs of Lake Turkana during the last five million years; there is still uncertainty about how typical this will prove to be.

The major claim made by the punctuationists is that the large-scale features of evolution—'macroevolution'—can be 'decoupled' from the processes occurring within populations that are studied in existing species by ecologists and population geneticists. Even if the minor claim proves to be correct (and as we have seen, it corresponds quite well with what Darwin expected), the major claim does not necessarily follow. Thus, if punctuational changes, when they occur, are caused by directional selection within populations, no decoupling exists. If, however, the characteristics of new species are random relative to the overall direction of macroevolution, and are the result of 'hopeful monsters' reproductively isolated from their ancestral species from the outset, decoupling would be real. As argued by Jones in his comment on Williamson's paper, the Lake Turkana molluscs support the minor claim, but there is no reason to think that any processes occurred other than those which can be studied in contemporary populations. The rapid changes, when they did happen, took perhaps 50,000 years to complete, and occurred in a population of many millions.

Williamson makes clear in his reply that he does not accept this interpretation. At the risk of being unfair, I shall now have, if not the last word, at least the last word here. First, a slightly peripheral point. Williamson is puzzled that the theory of punctuation and stasis should have been conflated with Goldschmidt's ideas about hopeful monsters. He is quite right to say that there is no necessary connection; as I have argued above, the conflation is an example of the unjustified leap from the minor to the major claim. However, if he reads the article by Gould, he will see that the leap is made by the punctuationists, not their critics.

Williamson's main points, however, are that neo-Darwinism did not predict and cannot explain stasis, and that the punctuational change involved a breakdown of a pre-existing 'developmental homeostasis', as evidenced by the increased variability of his transitional populations. Although there are other possible explanations of the increased variance, his suggestion about developmental homeostasis is interesting. However, the phenomenon—that the typical phenotype (in this context, shape) of a species is well buffered against both genetic and environmental change, whereas atypical phenotypes are poorly buffered— is a familiar one, discovered by population geneticists working mainly with laboratory populations of *Drosophila*. Further, homeostasis was thought (by Lerner, Waddington, Rendel and others who worked on it) to be brought about by 'stabilizing selection', that is, by selection favouring typical rather than extreme phenotypes.

How, then, is stasis to be explained? Williamson favours the view that it is maintained by 'developmental constraints', rather than by stabilizing selection. Developmental constraints, making certain kinds of phenotypic change difficult or impossible, certainly exist. However, they can hardly be responsible for stasis, because we know that selection can and does change populations, in nature and in captivity, in many directions. Stabilizing selection must also be involved. Indeed, if Lerner and Waddington were correct, stabilizing selection was itself responsible for at least some of the developmental constraints.

Hence I think that Williamson is wrong when he says that neo-Darwinism is unable to explain stasis, and that his results imply a decoupling between macro- and micro-evolution. However, he is clearly correct in saying that neo-Darwinism did not predict stasis. We must distinguish between the statements 'event X is incompatible with neo-Darwinist theory', and 'event X could not be predicted by that theory'. If the former is true, it is grounds for rejecting neo-Darwinism; if the latter, it indicates the need for further theoretical work. To give an example from physics, compare 'the orbit of Mercury is incompatible with

Newtonian mechanics' with 'the history of the solar system could not be predicted by Newtonian mechanics'. Both statements are true, but only the former provides grounds for rejecting Newton.

How should we respond to the conclusion that neo-Darwinism is compatible with what we know of macroevolution, but cannot predict it? Most biologists would assert that the major features of evolution (like those of history) are in the nature of things unpredictable, and can only be described and analysed in retrospect. However, there may be general features of the evolutionary process (for example, adaptive radiations; constant species diversity despite speciation and extinction) which can be established by a study of the fossil record, and which call for an explanation: hence the need for further theoretical work.

What kind of theory? It has been suggested that what is needed is an injection of developmental biology. I doubt this—much as I would love to see the formulation of adequate theories of development. If stasis, when it occurs, is to be explained by stabilising selection, and punctuation by directional selection, as Lande argues, what we need is a theory which says something about selection, and hence about the environment. Since the major component of the environment of most species consists of other species in the ecosystem, it follows that we need a theory of ecosystems in which the component species are evolving by natural selection. Such a theory is hard to formulate: the data needed to test it will be, primarily, the data of palaeontology.

The final paper in this section, by Boag and Grant, is less controversial (or so I thought until I saw it quoted in a major British newspaper as evidence against Darwinism). It is printed here as a particularly clear example of how a natural population can change under natural selection. Darwin would have been delighted with it, especially because it concerns the very finches that played such a crucial role in the genesis of his ideas.

Is a new and general theory of evolution emerging?

STEPHEN JAY GOULD

The modern synthesis, as an exclusive proposition, has broken down on both of its fundamental claims: extrapolationism (gradual allelic substitution as a model for all evolutionary change) and nearly exclusive reliance on selection leading to adaptation. Evolution is a hierarchical process with complementary, but different, modes of change at its three major levels: variation within populations, speciation, and patterns of macroevolution. Speciation is not always an extension of gradual, adaptive allelic substitution to greater effect, but may represent, as Goldschmidt argued, a different style of genetic change—rapid reorganization of the genome, perhaps non-adaptive. Macroevolutionary trends do not arise from the gradual, adaptive transformation of populations, but usually from a higher-order selection operating upon groups of species, while the individual species themselves generally do not change following their geologically instantaneous origin. I refer to these two discontinuities in the evolutionary hierarchy as the Goldschmidt break (between change in populations and speciation) and the Wright break (between speciation and trends as differential success among species).

A new and general evolutionary theory will embody this notion of hierarchy and stress a variety of themes either ignored or explicitly rejected by the modern synthesis: punctuational change at all levels, important non-adaptive change at all levels, control of evolution not only by selection, but equally by constraints of history, development and architecture—thus restoring to evolutionary theory a concept of organism.

THE MODERN SYNTHESIS

IN one of the last skeptical books written before the Darwinian tide of the modern synthesis asserted its hegemony, Robson and Richards[1] characterized the expanding orthodoxy that they deplored:

> The theory of Natural Selection . . . postulates that the evolutionary process is unitary, and that not only are groups formed by the multiplication of single variants having survival value, but also that such divergences are amplified to produce adaptations (both specializations

and organization). It has been customary to admit that certain ancillary processes are operative (isolation, correlation), but the importance of these, as active principles, is subordinate to selection (1936, pp. 370–371).

Darwinism, as a set of ideas, is sufficiently broad and variously defined to include a multitude of truths and sins. Darwin himself disavowed many interpretations made in his name (ref. 2, for example). The version known as the 'modern synthesis' or 'Neo-Darwinism' (different from what the late 19th century called Neo-Darwinism—see Romanes[3], 1900) is, I think, fairly characterized in its essentials by Robson and Richards[1]. Its foundation rests upon two major premises: (1) Point mutations (micromutations) are the ultimate source of variability. Evolutionary change is a process of gradual allelic substitution within a population. Events at broader scale, from the origin of new species to long-ranging evolutionary trends, represent the same process, extended in time and effect—large numbers of allelic substitutions incorporated sequentially over long periods of time. In short, gradualism, continuity and evolutionary change by the transformation of populations. (2) Genetic variation is raw material only. Natural selection directs evolutionary change. Rates and directions of change are controlled by selection with little constraint exerted by raw material (slow rates are due to weak selection, not insufficient variation). All genetic change is adaptive (though some phenotypic effects, due to pleiotropy, etc., may not be). In short, selection leading to adaptation.

All these statements, as Robson and Richards[1] also note, are subject to recognized exceptions—and this imposes a great frustration upon anyone who would characterize the modern synthesis in order to criticize it. All the synthesists recognized exceptions and 'ancillary processes,' but they attempted both to prescribe a low relative frequency for them and to limit their application to domains of little evolutionary importance. Thus, genetic drift certainly occurs—but only in populations so small and so near the brink that their rapid extinction will almost certainly ensue. And phenotypes include many non-adaptive features by allometry and pleiotropy, but all are epiphenomena of primarily adaptive genetic changes and none can have any marked effect upon the organism (for, if inadaptive, they will lead to negative selection and elimination and, if adaptive, will enter the model in their own right). Thus, a synthesist could always deny a charge of rigidity by invoking these official exceptions, even though their circumscription, both in frequency and effect, actually guaranteed the hegemony of the two cardinal principles. This frustrating situation had been noted by critics of an earlier Darwinian orthodoxy, by Romanes[3] writing of Wallace, for example (1900, p. 21):

[For Wallace,] the law of utility is, to all intents and purposes, universal, with the result that natural selection is virtually the only cause of organic evolution. I say 'to all intents and purposes,' or 'virtually,' because Mr. Wallace does not expressly maintain the abstract impossibility of laws and causes other than those of utility and natural selection; indeed, at the end of his treatise, he quotes with approval

Darwin's judgement, that 'natural selection has been the most important, but not the exclusive means of modification.' Nevertheless, as he nowhere recognizes any other law or cause of adaptive evolution, he practically concludes that, on inductive or empirical grounds, there *is* no such other law or cause to be entertained.

Lest anyone think that Robson and Richards, as doubters, had characterized the opposition unfairly, or that their two principles represent too simplistic or unsubtle a view of the synthetic theory, I cite the characterization of one of the architects of the theory himself (Mayr[4] 1963, p. 586—the first statement of his chapter on species and transspecific evolution):

> The proponents of the synthetic theory maintain that all evolution is due to the accumulation of small genetic changes, guided by natural selection, and that transspecific evolution is nothing but an extrapolation and magnification of the events that take place within populations and species.

The early classics of the modern synthesis—particularly Dobzhansky's[5] first edition (1937) and Simpson's[6] first book (1944)—were quite expansive, generous and pluralistic. But the synthesis hardened throughout the late 40's and 50's, and later editions of the same classics[7,8] (in 1951 and 1953) are more rigid in their insistence upon micromutation, gradual transformation and adaptation guided by selection (see Gould[9] for an analysis of changes between Simpson's two books). When Watson and Crick then determined the structure of DNA, and when the triplet code was cracked a few years later, everything seemed to fall even further into place. Chromosomes are long strings of triplets coding, in sequence, for the proteins that build organisms. Most point mutations are simple base substitutions. A physics and chemistry had been added, and it squared well with the prevailing orthodoxy.

I well remember how the synthetic theory beguiled me with its unifying power when I was a graduate student in the mid-1960's. Since then I have been watching it slowly unravel as a universal description of evolution. The molecular assault came first, followed quickly by renewed attention to unorthodox theories of speciation and by challenges at the level of macroevolution itself. I have been reluctant to admit it—since beguiling is often forever—but if Mayr's characterization of the synthetic theory is accurate, then that theory, as a general proposition, is effectively dead, despite its persistence as textbook orthodoxy.

REDUCTION AND HIERARCHY

The modern synthetic theory embodies a strong faith in reductionism. It advocates a smooth extrapolation across all levels and scales—from the base substitution to the origin of higher taxa. The most sophisticated of leading introductory textbooks in biology still proclaims:

> [Can] more extensive evolutionary change, macroevolution, be ex-

plained as an outcome of these microevolutionary shifts. Did birds really arise from reptiles by an accumulation of gene substitutions of the kind illustrated by the raspberry eye-color gene.

The answer is that it is entirely plausible, and no one has come up with a better explanation . . . . The fossil record suggests that macro-evolution is indeed gradual, paced at a rate that leads to the conclusion that it is based upon hundreds or thousands of gene substitutions no different in kind from the ones examined in our case histories (Wilson et al.[10] pp. 793–794).

The general alternative to such reductionism is a concept of hierarchy—a world constructed not as a smooth and seamless continuum, permitting simple extrapolation from the lowest level to the highest, but as a series of ascending levels, each bound to the one below it in some ways and independent in others. Discontinuities and seams characterize the transitions; 'emergent' features not implicit in the operation of processes at lower levels, may control events at higher levels. The basic processes—mutation, selection, etc.—may enter into explanations at all scales (and in that sense we may still hope for a general theory of evolution), but they work in different ways on the characteristic material of divers levels (see Bateson[11] and Koestler[12], for all its other inadequacies, for good discussions of hierarchy and its anti-reductionistic implications; Eldredge and Cracraft[13]).

The molecular level, which once seemed through its central dogma and triplet code to provide an excellent 'atomic' basis for smooth extrapolation, now demands hierarchical interpretation itself. The triplet code is only machine language (I thank E. Yates for this appropriate metaphor). The program resides at a higher level of control and regulation—and we know virtually nothing about it. With its inserted sequences and jumping genes, the genome contains sets of scissors and pots of glue to snip and unite bits and pieces from various sources. Thirty to seventy percent of the mammalian genome consists of repetitive sequences, some repeated hundreds or thousands of times. What are they for (if anything)? What role do they play in the regulation of development? Molecular biologists are groping to understand this higher control upon primary products of the triplet code. In that understanding, we will probably obtain a basis for styles of evolutionary change radically different from the sequential allelic substitutions, each of minute effect, that the modern synthesis so strongly advocated. The uncovering of hierarchy on the molecular level will probably exclude smooth continuity across other levels. (We may find, for example, that structural gene substitutions control most small-scale, adaptive variation within local populations, while disruption of regulation lies behind most key innovations in macroevolution.)

The modern synthesis drew most of its direct conclusions from studies of local populations and their immediate adaptations. It then extrapolated the postulated mechanism of these adaptations—gradual, allelic substitution—to encompass all larger-scale events. The synthesis is now breaking down on both sides of this argument. Many evolutionists now doubt exclusive control by selection upon genetic change within local populations. Moreover, even if local populations alter as the synthesis maintains, we now doubt that the same style of change controls events at the two major higher levels: speciation and patterns of macroevolution.

A NOTE ON LOCAL POPULATIONS AND NEUTRALITY

At the level of populations, the synthesis has broken on the issue of amounts of genetic variation. Selection, though it eliminates variation in both its classical modes (directional and, especially, stabilizing) can also act to preserve variation through such phenomena as overdominance, frequency dependence, and response to small-scale fluctuation of spatial and temporal environments. Nonetheless, the copiousness of genetic variation, as revealed first in the electrophoretic techniques that resolve only some of it[14,15] cannot be encompassed by our models of selective control (of course, the models, rather than nature, may be wrong). This fact has forced many evolutionists, once stout synthesists themselves, to embrace the idea that alleles often drift to high frequency or fixation, and that many common variants are therefore neutral or just slightly deleterious. This admission lends support to a previous interpretation of the approximately even ticking of the molecular clock[16]—that it reflects the neutral status of most changes in structural genes rather than a grand averaging of various types of selection over time.

None of this evidence, of course, negates the role of conventional selecton and adaptation in molding parts of the phenotype with obvious importance for survival and reproduction. Still, it rather damps Mayr's enthusiastic claim for 'all evolution . . . guided by natural selection.' The question, as with so many issues in the complex sciences of natural history, becomes one of relative frequency. Are the Darwinian substitutions merely a surface skin on a sea of variation invisible to selection, or are the neutral substitutions merely a thin bottom layer underlying a Darwinian ocean above? Or where in between?

In short, the specter of stochasticity has intruded upon explanations of evolutionary *change*. This represents a fundamental challenge to Darwinism, which holds, as its very basis, that random factors enter only in the production of raw material, and that the deterministic process of selection produces change and direction (see ref. 17).

THE LEVEL OF SPECIATION AND THE GOLDSCHMIDT BREAK

Even since Darwin called his book *The Origin of Species*, evolutionists have regarded the formation of reproductively isolated units by speciation as a fundamental process of large-scale change. Yet speciation occurs at too high a level to be observed directly in nature or produced by experiment in most cases. Therefore, theories of speciation have been based on analogy, extrapolation and inference. Darwin himself focused on artificial selection and geographic variation. He regarded subspecies as incipient species and viewed their gradual, accumulating divergence as the primary mode of origin for new taxa. The modern synthesis continued this tradition of extrapolation from local populations and used the accepted model for adaptive geographic variation—gradual allelic substitution directed by natural selection—as a paradigm for the origin of species. Mayr's[4, 18] model of allopatric speciation did challenge Darwin's implied notion of sympatric continuity. It emphasized the crucial role of isolation from gene flow and did promote the importance of small founding populations and relatively rapid rates of change. Thus, the small peripheral isolate, rather than the large local population in persistent contact with other conspecifics, became the incipient species. Nonetheless, despite this welcome departure from the purest form of Darwinian gradualism, the allopatric theory held firmly to the two major principles that permit smooth extrapolation from the *Biston betularia* model of adaptive, allelic substitution: (i) The accumulating changes that lead to speciation are adaptive. Reproductive isolation is a consequence of sufficient accumulation. (ii) Although aided by founder effects and even (possibly) by drift, although dependent upon isolation from gene flow, although proceeding more rapidly than local differentiation within large populations, successful speciation is still a cumulative and sequential process powered by selection through large numbers of generations. It is, if you will, Darwinism a little faster.

I have no doubt that many species originate in this way; but it now appears that many, perhaps most, do not. The new models stand at variance with the synthetic proposition that speciation is an extension of microevolution within local populations. Some of the new models call upon genetic variation of a different kind, and they regard reproductive isolation as potentially primary and non-adaptive rather than secondary and adaptive. Insofar as these new models be valid in theory and numerically important in application, speciation is not a simple 'conversion' to larger effect of processes occurring at the lower level of adaptive modelling within local populations. It represents a discontinuity in our hierarchy of explanations, as the much maligned Richard Goldschmidt argued explicitly in 1940 (ref. 19).

There are many ways to synthesize the swirling set of apparently disparate challenges that have rocked the allopatric orthodoxy and established an alternative set of models for speciation. The following reconstruction is neither historically sequential nor the only logical pathway of linkage, but it does summarize the challenges—on population structure, place of origin, genetic style, rate, and relation to adaptation—in some reasonable order.

1. Under the allopatric orthodoxy, species are viewed as integrated units which, if not actually panmictic, are at least sufficiently homogenized by gene flow to be treated as entities. This belief in effective homogenization within central populations underlies the allopatric theory with its emphasis on *peripheral* isolation as a precondition for speciation. But many evolutionists now believe that gene flow is often too weak to overcome selection and other intrinsic processes within local demes[20]. Thus, the model of a large, homogenized central population preventing local differentiation and requiring allopatric 'flight' of isolated

demes for speciation may not be generally valid. Perhaps most local demes have the required independence for potential speciation.

2. The primary terms of reference for theories of speciation—allopatry and sympatry—lose their meaning if we accept the first statement. Objections to sympatric speciation centered upon the homogenizing force of gene flow. But if demes may be independent in all geographic domains of a species, then sympatry loses its meaning and allopatry its necessity. Independent demes within the central range (sympatric by location) function, in their freedom from gene flow, like the peripheral isolates of allopatric theory. In other words, the terms make no sense outside a theory of population structure that contrasts central panmixia with marginal isolation. They should be abandoned.

3. In this context 'sympatric' speciation loses its status as an extremely improbable event. If demes are largely independent, new species may originate anywhere within the geographic range of an ancestral form. Moreover, many evolutionists now doubt that parapatric distributions (far more common then previously thought) must represent cases of secondary contact. White[21] (p. 342) believes that many, if not most, are primary and that speciation can also occur between populations continually in contact if gene flow can be overcome either by strong selection or by the sheer rapidity of potential fixation for major chromosomal variants (see White[21], p. 17 on clinal speciation).

4. Most 'sympatric' models of speciation are based upon rates and styles of genetic change inconsistent with the reliance placed by the modern synthesis on slow, or at least sequential change.

The most exciting entry among punctuational models for speciation in ecological time is the emphasis, now coming from several quarters, on chromosomal alterations as isolating mechanisms[21-25]—sometimes called the theory of chromosomal speciation. In certain population structures, particularly in very small and circumscribed groups with high degrees of inbreeding, major chromosomal changes can rise to fixation in less than a handful of generations (mating of heterozygous F_1 sibs to produce F_2 homozygotes for a start).

Allan Wilson, Guy Bush and their colleagues[24,25] find a strong correlation between rates of karyotypic and anatomical change, but no relation between amounts of substitution in structural genes and any conventional assessment of phenotypic modification, either in speed or extent. They suggest that speciation may be more a matter of gene regulation and rearrangement than of changes in structural genes that adapt local populations in minor ways to fluctuating environments (the *Biston betularia* model).

Carson[23,26] has also stressed the importance of small demes, chromosomal change, and extremely rapid speciation in his founder-flush theory with its emphasis on extreme bottlenecking during crashes of the flush-crash cycle (see Powell[27] for experimental support). Explicitly contrasting this view with extrapolationist models based on sequential substitution of structural genes, he writes (ref. 23, p. 88):

Most theories of speciation are wedded to gradualism, using the mode of origin of intraspecific adaptations as a model . . . I would nevertheless like to propose . . . that speciational events may be set in motion and important genetic saltations towards species formation accomplished by a series of catastrophic, stochastic genetic events . . . initiated when an unusual forced reorganization of the epistatic supergenes of the closed variability system occurs . . . I propose that this cycle of disorganization and reorganization be viewed as the essence of the speciation process.

5. Another consequence of such essentially saltational origin is even more disturbing to conventional views than the rapidity of the process itself, as Carson has forcefully stated. The control of evolution by selection leading to adaptation lies at the heart of the modern synthesis. Thus, reproductive isolation, the definition of speciation, is attained as a by-product of adaptation—that is, a population diverges by sequential adaptation and eventually becomes sufficiently different from its ancestor to foreclose interbreeding. (Selection for reproductive isolation may also be direct when two imperfectly-separate forms come into contact.) But in saltational, chromosomal speciation, reproductive isolation comes first and cannot be considered as an adaptation at all. It is a stochastic event that establishes a species by the technical definition of reproductive isolation. To be sure, the later success of this species in competition may depend upon its subsequent acquisition of adaptations; but the origin itself may be non-adaptive. We can, in fact, reverse the conventional view and argue that speciation, by forming new entities stochastically, provides raw material for selection.

These challenges can be summarized in the claim that a discontinuity in explanation exists between allelic substitutions in local populations (sequential, slow and adaptive) and the origin of new species (often discontinuous and non-adaptive). During the heyday of the modern synthesis, Richard Goldschmidt was castigated for his defense of punctuational speciation. I was told as a graduate student that this great geneticist has gone astray because he had been a lab man with no feel for nature, a person who hadn't studied the adaptation of local populations and couldn't appreciate its potential power, by extrapolation, to form new species. But I discovered, in writing *Ontogeny and Phylogeny*, that Goldschmidt had spent a good part of his career studying geographic variation, largely in the coloration of lepidopteran larvae (where he developed the concept of rate genes to explain minor changes in pattern). I then turned to his major book[19] and found that his defense of saltational speciation is not based on ignorance of geographic variation, but on an explicit study of it; half the book is devoted to this subject. Goldschmidt concludes that geographic variation is ubiquitous, adaptive, and essential for the persistence of established species. But it is simply not the stuff of speciation; it is a different process. Speciation, Goldschmidt argues, occurs at different rates and uses different kinds of genetic variation. We do not now accept all his arguments about the nature of variation, but his explicit anti-extrapolationist statement is the epitome and foundation of emerging views on speciation discussed in this section. There is a discontinuity in cause and explanation between adaptation in local populations and speciation; they represent two distinct, though interacting, levels of evolution. We might refer to this discontinuity as the *Goldschmidt break*, for he wrote:

> The characters of subspecies are of a gradient type, the species limit is characterized by a gap, an unbridged difference in many characters. This gap cannot be bridged by theoretically continuing the subspecific gradient or cline beyond its actually existing limits. The subspecies do not merge into the species either actually or ideally . . . . Micro-evolution by accumulation of micromutations—we may also say neo-Darwinian evolution—is a process which leads to diversification strictly within the species, usually, if not exclusively, for the sake of adaptation of the species to specific conditions within the area which it is able to occupy . . . . Subspecies are actually, therefore, neither incipient species nor models for the origin of species. They are more or less diversified blind alleys within the species. The decisive step in evolution, the first step towards macroevolution, the step from one species to another, requires another evolutionary method than that of sheer accumulation of micromutations (ref. 19, p. 183).

MACROEVOLUTION AND THE WRIGHT BREAK

The extrapolationist model of macroevolution views trends and major transitions as an extension of allelic substitution within populations—the march of frequency distributions through time. Gradual change becomes the normal state of species. The discontinuities of the fossil record are all attributed to its notorious imperfection; the remarkable stasis exhibited by most species during millions of years is ignored (as no data), or relegated to descriptive sections of taxonomic monographs. But gradualism is not the only important implication of the extrapolationist model. Two additional consequences have channeled our concept of macroevolution, both rather rigidly and with unfortunate effect. First, the trends and transitions of macroevolution are envisaged as events in the phyletic mode—populations transforming themselves steadily through time. Splitting and branching are acknowledged to be sure, lest life be terminated by its prevalent extinctions. But splitting becomes a device for the generation of diversity upon designs attained through 'progressive' processes of transformation. Splitting, or cladogenesis, becomes subordinate in importance to transformation, or anagenesis (see Ayala[28], p. 141; but also see Mayr[4], p. 621 for a rather lonely voice in the defense of copious speciation as an input to 'progressive' evolution). Secondly, the adaptationism that prevails in interpreting change in local populations gains greater confidence in extrapolation. For if allelic substitutions in ecological time have an adaptive basis, then surely a unidirectional trend that persists for millions of years within a single lineage cannot bear any other interpretation.

This extrapolationist model of adaptive, phyletic gradualism has been vigorously challenged by several paleobiologists—and again with a claim for discontinuity in explanation at different levels. The general challenge embodies three loosely united themes:

1. Evolutionary trends as a higher level process: Eldredge and I have argued[29,30] that imperfections of the record cannot explain all discontinuity (and certainly cannot encompass stasis). We regard stasis and discontinuity as an expression of how evolution works when translated into geological time. Gradual change is not the normal state of a species. Large, successful central populations undergo minor adaptive modifications of fluctuating effect through time (Goldschmidt's 'diversified blind alleys'), but they will rarely transform *in toto* to something fundamentally new. Speciation, the basis of macroevolution, is a process of branching. And this branching, under any current model of speciation—conventional allopatry to chromosomal saltation—is so rapid in geological translation (thousands of years at most compared with millions for the duration of most fossil species) that its results should generally lie on a bedding plane, not through the thick sedimentary sequence of a long hillslope. (The expectation of gradualism emerges as a kind of double illusion. It represents, first of all, an incorrect translation of conventional allopatry. Allopatric speciation seems so slow and gradual in ecological time that most paleontologists never recognized it as a challenge to the style of gradualism—steady change over millions of years—promulgated by custom as a model for the history of life. But it now appears that 'slow' allopatry itself may be less important than a host of alternatives that yield new species rapidly even in ecological time.) Thus, our model of 'punctuated equilibria' holds that evolution is concentrated in events of speciation and that successful speciation is an infrequent event punctuating the stasis of large populations that do not alter in fundamental ways during the millions of years that they endure.

But if species originate in geological instants and then do not alter in major ways, then evolutionary trends cannot represent a simple extrapolation of allelic substitution within a population. Trends must be the product of differential success among species[29,31]. In other words, species themselves must be inputs, and trends the result of their differential origin and survival. Speciation interposes itself as an irreducible level between change in local populations and trends in geological time. Macroevolution is, as Stanley argues[31] (p. 648), decoupled from microevolution.

Sewall Wright recognized the hierarchical implications of viewing species as irreducible inputs to macroevolution when he argued[32] (p. 121) that the relationship between change in local populations and evolutionary trends can only be analogical. Just as mutation is random with respect to the direction of change within a population, so too might speciation be random with respect to the direction of a macroevolutionary trend. A higher form of selection, acting directly upon species through differential rates of extinction, may then be the analog of natural selection working within populations through differential mortality of individuals.

Evolutionary trends therefore represent a third level superposed upon speciation and change within demes. Intrademic events cannot encompass speciation because rates, genetic styles, and relation to adaptation differ for the two processes. Likewise, since trends 'use' species as their raw material, they represent a process at a higher level than speciation itself. They reflect a sorting out of speciation events. With apologies for the pun, the hierarchical rupture between speciation and macroevolutionary trends might be called the Wright break.*

As a final point about the extrapolation of methods for the study of events within populations, the cladogenetic basis of macroevolution virtually precludes any direct application of the primary apparatus for microevolutionary theory: classical population genetics. I believe that essentially all macroevolution is cladogenesis and its concatenated effects. What we call 'anagenesis' and often attempt to delineate as a separate phyletic process leading to 'progress,' is just accumulated cladogenesis filtered through the directing force of species selection[31]— Wright's higher level analog of natural selection. Carson[26] (p. 925) makes the point forcefully, again recognizing Sewall Wright as its long and chief defender:

> Investigation of cladistic events as opposed to phyletic (anagenetic) ones requires a different perspective from that normally assumed in classical population genetics. The statistical and mathematical comfort of the Hardy-Weinberg equilibrium in large populations has to be abandoned in favour of the vague realization that nearly everywhere in nature we are faced with data suggesting the partial or indeed complete sundering of gene pools. If we are to deal realistically with cladogenesis we must seek to delineate each genetic and environmental factor which may promote isolation. The most important devices are clearly those which operate at the very lowest population level: sib from sib, family from family, deme from deme. Formal population genetics just cannot deal with such things, as Wright pointed out long ago.

Eldredge[33] has traced many conceptual errors and prejudicial blockages to our tendency for conceiving of evolution as the transformation of *characters* within phyletic lineages, rather than as the origin of new *taxa* by cladogenesis (the transformational versus the taxic view in his terms). I believe that, in ways deeper than we realize, our preference for transformational thinking represents a cultural tie to the controlling Western

* I had the honor—not a word I use frequently, but inescapable in this case—of spending a long evening with Dr. Wright last year. I discovered that his quip about macroevolution, just paraphrased, was no throwaway statement but an embodiment of his deep commitment to a hierarchical view of evolutionary causation. (The failure of many evolutionists to think hierarchically is responsible for the most frequent misinterpretation of Wright's views. He never believed that genetic drift—the Sewall Wright effect as it once was called—is an important agent of evolutionary *change*. He regards it as input to the directional process of interdemic selection for evolution within species. Drift can push a deme off an adaptive peak; selection can then draw it to another peak.

themes of progress and ranking by intrinsic merit—an attitude that can be traced in evolutionary thought to Lamarck's distinction between the march up life's ladder promoted by the *pouvoir de la vie* and the tangential departures imposed by *l'influence des circonstances*, with the first process essential and the second deflective. Nonetheless, macroevolution is fundamentally about the origin of taxa by splitting.

2. The saltational initiation of major transitions: The absence of fossil evidence for intermediary stages between major transitions in organic design, indeed our inability, even in our imagination, to construct functional intermediates in many cases, has been a persistent and nagging problem for gradualistic accounts of evolution. In 1871 St. George Mivart[34], Darwin's most cogent critic, referred to it as the dilemma of 'the incipient stages of useful structures'—of what possible benefit to a reptile is two percent of a wing? The dilemma has two potential solutions. The first, preferred by Darwinians because it preserves both gradualism and adaptation, is the principle of preadaptation: the intermediary stages functioned in another way but were, by good fortune in retrospect, pre-adapted to a new role they could play only after greater elaboration. Thus, if feathers first functioned 'for' insulation and later 'for' the trapping of insect prey[35], a proto-wing might be built without any reference to flight.

I do not doubt the supreme importance of preadaptation, but the other alternative, treated with caution, reluctance, disdain or even fear by the modern synthesis, now deserves a rehearing in the light of renewed interest in development: perhaps, in many cases, the intermediates never existed. I do not refer to the saltational origin of entire new designs, complete in all their complex and integrated features—a fantasy that would be truly anti-Darwinian in denying any creativity to selection and relegating it to the role of eliminating old models. Instead, I envisage a potential saltational origin for the essential features of key adaptations. Why may we not imagine that gill arch bones of an ancestral agnathan moved forward in one step to surround the mouth and form proto-jaws? Such a change would scarcely establish the *Bauplan* of the gnathostomes. So much more must be altered in the reconstruction of agnathan design—the building of a true shoulder girdle with bony, paired appendages, to say the least. But the discontinuous origin of a proto-jaw might set up new regimes of development and selection that would quickly lead to other, coordinated modifications. Yet Darwin, conflating gradualism with natural selection as he did so often, wrongly proclaimed that any such discontinuity, even for organs (much less taxa) would destroy his theory:

> If it could be demonstrated that any complex organ existed, which could not possibly have been formed by numerous, successive, slight modifications, my theory would absolutely break down[36] (p. 189).

During the past 30 years, such proposals have generally been treated as a fantasy signifying surrender—an invocation of hopeful monsters rather than a square facing of a difficult issue. But our renewed interest in development, the only discipline of biology that might unify molecular and evolutionary approaches into a coherent science, suggests that such ideas are neither fantastic, utterly contrary to genetic principles, nor untestable.

Goldschmidt conflated two proposals as causes for hopeful monsters—'systemic mutations' involving the entire genome (a spinoff from his fallacious belief that the entire genome acted as an integrated unit), and small mutations with large impact upon adult phenotypes because they work upon early stages of ontogeny and lead to cascading effects throughout embryology. We reject his first proposal, but the second, eminently plausible, theme might unite a Darwinian insistence upon continuity of genetic change with a macroevolutionary suspicion of phenetic discontinuity. It is, after all, a major focus in the study of heterochrony (effects, often profound, of small changes in developmental rate upon adult phenotypes); it is also implied in the emphasis now being placed upon regulatory genes in the genesis of macroevolutionary change[37]—for regulation is fundamentally about timing in the complex orchestration of development. Moreover, although we cannot readily build 'hopeful monsters,' the subject of major change through alteration of developmental rate can be treated, perhaps more than analogically, both by experiment and comparative biology. The study of spontaneous anomalies of development (teratology) and experimental perturbations of embryogenic rates explores the tendencies and boundaries of developmental systems and allows us to specify potential pathways of macroevolutionary change (see, for example, the stunning experiment of Hampé[38] on recreation of reptilian patterns in birds, after 200 million years of their phenotypic absence, by experimental manipulations that amount to alterations in rate of development for the fibula). At the very least, these approaches work with real information and seem so much more fruitful than the construction of adaptive stories or the invention of hypothetical intermediates.

3. The importance of non-adaptation: The emphasis on natural selection as the only directing force of any importance in evolution led inevitably to an analysis of all attributes of organisms as adaptations. Indeed, the tendency has infected our language, for, without thinking about what it implies, we use 'adaptation' as our favored, *descriptive* term for designating any recognizable bit of changed morphology in evolution. I believe that this 'adaptationist program' has had decidedly unfortunate effects in biology[39]. It has led to a reliance on speculative storytelling in preference to the analysis of form and its constraints; and, if wrong, in any case, it is virtually impossible to dislodge because the failure of one story leads to invention of another rather than abandonment of the enterprise.

Yet, as I argued earlier, the hegemony of adaptation has been broken at the two lower levels of our evolutionary hierarchy: variation within populations, and speciation. Most populations may contain too much variation for selection to maintain; moreover, if the neutralists are even part right, much allelic substitution occurs without controlling influence from selection, and with no direct relationship to adaptation. If species often form as a result of major chromosomal alterations, then their origin—the establishment of reproductive isolation—may require no reference to adaptation. Similarly, at this third level of macroevolution, both arguments previously cited against the conventional extrapolationist view require that we abandon strict adaptationism.

(i) If trends are produced by the unidirectional transformation of populations (orthoselection), then they can scarcely receive other than a conventional adaptive explanation. After all, if adaptation lies behind single allelic substitutions in the *Biston betularia* model for change in local populations, what else but even stronger, more persistent selection and adaptive orientation can render a trend that persists for millions of years? But if trends represent a higher-level process of differential origin and mortality among species, then a suite of potentially non-adaptive explanations must be considered. Trends, for example, may occur because some kinds of species tend to speciate more often than others. This tendency may reside in the character of environments or in attributes of behavior and population structure bearing no relationship to morphologies that spread through lineages as a result of higher speciation rates among some of their members. Or trends may arise from the greater longevity of certain kinds of species. Again, this greater persistence may have little to do with the morphologies that come to prevail as a result. I suspect that many morphological trends in paleontology—a bugbear of the profession because we have been unable to explain them in ordinary adaptive terms—are non-adaptive sequelae of differential species success based upon environments and population structures.

(ii) If transitions represent the continuous and gradual transformation of populations, then they must be regulated by adaptation throughout (even though adaptive orientation may alter according to the principle of preadaptation). But if discontinuity arises through shifts in development, then directions of potential change may be limited and strongly constrained by the inherited program and developmental mechanics of an organism. Adaptation may determine whether or not a hopeful monster survives, but primary constraint upon its genesis and direction resides with inherited ontogeny, not with selective modelling.

QUO VADIS?

My crystal ball is clouded both by the dust of these growing controversies and by the mists of ignorance emanating from molecular biology, where even the basis of regulation in eukaryotes remains shrouded in mystery. I think I can see what is breaking down in evolutionary theory—the strict construction of the modern synthesis with its belief in pervasive adaptation, gradualism, and extrapolation by smooth continuity from causes of change in local populations to major trends and transitions in the history of life. I do not know what will take its place as a unified theory, but I would venture to predict some themes and outlines.

The new theory will be rooted in a hierarchical view of nature. It will not embody the depressing notion that levels are fundamentally distinct and necessarily opposed to each other in their identification of causes (as the older paleontologists held in maintaining that macroevolution could not, in principle, be referred to the same causes that regulate micro-evolution—e.g. Osborn[40] 1922). It will possess a common body of causes and constraints, but will recognize that they work in characteristically different ways upon the material of different levels—intrademic change, speciation, and patterns of macroevolution.

As its second major departure from current orthodoxy, the new theory will restore to biology a concept of organism. In an exceedingly curious and unconscious bit of irony, strict selectionism (which was not, please remember, Darwin's own view) debased what had been a mainstay of biology—the organism as an integrated entity exerting constraint over its history. St. George Mivart expressed the subtle point well in borrowing a metaphor from Galton. I shall call it Galton's polyhedron. Mivart[34] writes (1871, pp. 228–229):

> This conception of such internal and latent capabilities is somewhat like that of Mr. Galton . . . according to which the organic world consists of entities, each of which is, as it were, a spheroid with many facets on its surface, upon one of which it reposes in stable equilibrium. When by the accumulated action of incident forces this equilibrium is disturbed, the spheroid is supposed to turn over until it settles on an adjacent facet once more in stable equilibrium. The internal tendency of an organism to certain considerable and definite changes would correspond to the facets on the surface of the spheroid.

Under strict selectionism, the organism is a sphere. It exerts little constraint upon the character of its potential change; it can roll along all paths. Genetic variation is copious, small in its increments, and available in all directions—the essence of the term 'random' as used to guarantee that variation serves as raw material only and that selection controls the direction of evolution.

By invoking Galton's polyhedron, I recommend no return to the antiquated and anti-Darwinian view that mysterious 'internal' factors provide direction inherently, and that selection only eliminates the unfit (orthogenesis, various forms of vitalism and finalism). Instead, the facets are constraints exerted by the developmental integration of organisms themselves. Change cannot occur in all directions, or with any incre-ment; the organism is not a metaphorical sphere. When the polyhedron tumbles, selection may usually be the propelling force. But if adjacent facets are few in number and wide in spacing, then we cannot identify selection as the only, or even the primary control upon evolution. For selection is channeled by the form of the polyhedron it pushes, and these constraints may exert a more powerful influence upon evolutionary directions than the external push itself. This is the legitimate sense of a much maligned claim that 'internal factors' are important in evolution. They channel and constrain Darwinian forces; they do not stand in opposition to them. Most of the other changes in evolutionary viewpoint that I have advocated throughout this paper fall out of Galton's metaphor: punctuational change at all levels (the flip from facet to facet, since homeostatic systems change by abrupt shifting to new equilibria);

essential non-adaptation, even in major parts of the phenotype (change in an integrated organism often has effects that reverberate throughout the system); channeling of direction by constraints of history and developmental architecture. Organisms are not billiard balls, struck in deterministic fashion by the cue of natural selection, and rolling to optimal positions on life's table. They influence their own destiny in interesting, complex, and comprehensible ways. We must put this concept of organism back into evolutionary biology.

1. Robson, G. C. & Richards, O. W. *The Variation of Animals in Nature.* (Longmans, Green, and Co.; London, 1936).

2. Darwin, C. Sir Wyville Thomson and natural selection. *Nature* 23, 32. (1880).

3. Romanes, G. J. *Darwin and after Darwin,* vol. 2. *Post-Darwinian questions. Heredity and Utility,* 344 pp. (Longmans, Green, and Co.; London, 1900).

4. Mayr, E. *Animal Species and Evolution.* 797 pp. (Belknap Press of Harvard Univ. Press; Cambridge, Mass., 1963).

5. Dobzhansky, Th. *Genetics and the Origin of Species.* 364 pp. (Columbia Univ. Press: New York, 1937).

6. Simpson, G. G. *Tempo and Mode in Evolution.* 237 pp. (Columbia Univ. Press; New York, 1944).

7. Dobzhansky, Th. *Genetics and the Origin of Species.* (3rd ed.). 364 pp. (Columbia Univ. Press; New York, 1951).

8. Simpson, G. G. *The Major Features of Evolution.* 434 pp. (Columbia Univ. Press; New York, 1953).

9. Gould, S. J. G. G. Simpson, paleontology and the modern synthesis. In: Mayr, E., ed. *Conference on the making of the Modern Synthesis.* (Harvard Univ. Press; Cambridge, Mass., 1980).

10. Wilson, E. O. *et al. Life on Earth.* (Sinauer Associates; Sunderland, Mass., 1973).

11. Bateson, G. *Mind and Nature.* (E. P. Dutton; New York, 1978).

12. Koestler, A. *Janus: a Summing Up.* (Random House; New York, 1978).

13. Eldredge, N. & Cracraft, J. *Phylogenetic Patterns and the Evolutionary Process.* (Columbia Univ. Press; New York, 1980).

14. Lewontin, R. C. & Hubby, J. L. A molecular approach to the study of genic heterozygosity in natural populations. II. Amount of variation and degree of heterozygosity in natural populations of *Drosophila pseudoobscura. Genetics.* 54, 595–609 (1966).

15. Lewontin, R. C. *The Genetic Basis of Evolutionary Change.* 346 pp. (Columbia Univ. Press; New York, 1974).

16. Wilson, A. C., Carlson, S. S. & White T. J. Biochemical evolution. *Annu. Rev. Biochem.* 46, 573–639 (1977).

17. Nei, M. *Molecular Population Genetics and Evolution.* (American Elsevier; New York, 1975).

18. Mayr, E. *Systematics and the Origin of Species.* 334 pp. (Columbia Univ. Press; New York, 1942).

19. Goldschmidt, R. *The Material Basis of Evolution.* 436 pp. (Yale Univ. Press; New Haven, Conn., 1940).

20. Ehrlich, P. R. & Raven, P. H. Differentiation of populations. *Science.* 165, 1228–1232 (1969).

21. White, M. J. D. *Modes of Speciation.* 455 pp. (W. H. Freeman; San Francisco, California, 1978).

22. Bush, G. L. Modes of animal speciation. *Annu. Rev. Ecol. Syst.* 6, 339–364 (1975).

23. Carson, H. L. The genetics of speciation at the diploid level. *Am. Nat.* 109, 83–92 (1975).

24. Wilson, A. C., Bush, G. L., Case, S. M. & King, M. C. Social structuring of mammalian populations and rate of chromosomal evolution. *Proc. Nat. Acad. Sci.* 72, 5061–5065 (1975).

25. Bush, G. L., Case, S. M., Wilson, A. C. & Patton, J. L. Rapid speciation and chromosomal evolution in mammals. *Proc. Nat. Acad. Sci.* 74, 3942–3946 (1977).

26. Carson, H. L. Chromosomes and species formation. *Evolution.* 32, 925–927 (1978).

27. Powell, J. R. The founder-flush speciation theory: an experimental approach. *Evolution.* 32, 465–474 (1978).

28. Ayala, F. J. Molecular genetics and evolution. Pp. 1–20. In: Ayala, F. J., ed. *Molecular Evolution.* (Sinauer Associates; Sunderland, Mass., 1976).

29. Eldredge, N. & Gould, S. J. Punctuated equilibria: An alternative to phyletic gradualism. Pp. 82–115. In: Schopf, T. J. M., ed. *Models in Paleobiology.* (Freeman, Cooper and Co.; San Francisco, California, 1972).

30. Gould, S. J. & Eldredge, N. Punctuated equilibria: the tempo and mode of evolution reconsidered. *Paleobiology.* 3, 115–151 (1977).

31. Stanley, S. M. A theory of evolution above the species level. *Proc. Nat. Acad. Sci.* 72, 646–650 (1975).

32. Wright, S. Comments on the preliminary working papers of Eden and Waddington. In: Moorehead, P. S. & M. M. Kaplan, eds. *Mathematical Challenges to the Neo-Darwinian Theory of Evolution. Wistar Inst. Symp.* 5, 117–120 (1967).

33. Eldredge, N. Alternative approaches to evolutionary theory. *Bull. Carnegie Mus. Nat. Hist.* pp. 7–19 (1979).

34. Mivart, St. G. *On the Genesis of Species.* 296 pp. (MacMillan; London, 1871).

35. Ostrom, J. H. Bird flight: How did it begin. *Am. Sci.* 67, 46–56 (1979).

36. Darwin, C. *On the Origin of Species.* 490 pp. (John Murray; London, 1859).

37. King, M. C. & Wilson, A. C. Evolution at two

levels in humans and chimpanzees. *Science.* **188**, 107–116 (1975).

38. Hampé, A. Contribution à l'étude du développement et de la regulation des déficiences et des excédents dans la patte de l'embroyon de poulet. *Arch. Anat. Microsc. Morphol. Exp.* **48**, 345–478 (1959).

39. Gould, S. J. & Lewontin, R. C. The spandrels of San Marco and the Panglossian paradigm: A critique of the adaptationist program. *Proc. R. Soc. London.* **205**, 581–598 (1979).

40. Osborn, H. F. Orthogenesis as observed from paleontological evidence beginning in the year 1889. *Am. Nat.* **56**, 134–143 (1922).

Stephen Jay Gould is at the Museum of Comparative Zoology, Harvard University.

This article was first published in *Paleobiology* Vol. **6**(1), pp. 119–130; 1980. It is reproduced with permission.

Microevolution in relation to macroevolution

RUSSELL LANDE

*Macroevolution** (ref. 1) is the most comprehensive text written by a
paleontologist since Simpson's *The Major Features of Evolution*[2] (1953). In
ten chapters Stanley covers the topics of speciation mechanisms, rates of
extinction and speciation, and the controls of organic diversity. A
wealth of facts from functional morphology, ecology, paleontology and
genetics is marshalled to support the two main ideas of the book,
'punctuated equilibrium': that most morphological change occurs during
rapid speciation in small populations, and 'species selection': that most
large-scale patterns and trends result from differential extinction and
origination of phenotypically stable species. Stanley summarizes recent
work on these controversial subjects and presents many new examples
and statistics on rates of taxonomic evolution.

A major thesis of *Macroevolution* is that the fossil record is not as
incomplete as formerly believed and that rapid speciation may result
from changes in a small number of regulatory mutations acting early in
development, thus explaining many of the discontinuities between living
and extinct species. The recurrent popularity of macromutations and
hopeful monsters seems to rely more on their ready explanation for
apparently sudden morphological changes than on any compelling
evidence in their favor, since the available genetic information indicates
that successful macromutations are extremely rare. Most modern
authors treat selection at the levels of the individual and of the species as
potentially opposing processes in which the balance may be tipped either
way by various factors. Stanley argues that over a sufficiently long time
span species selection will generally predominate. The importance of
these topics justifies their critical reexamination from many points of
view. For brevity, the following remarks are restricted to genetic aspects
of speciation, and rates and patterns of phenotypic evolution. Many of
these remarks also apply to works by other authors as indicated.

Macroevolution: Pattern and Process. Steven M. Stanley. W. H. Freeman
and Co.; San Francisco; 1979.

GENETIC ASPECTS OF SPECIATION

Eldredge and Gould[3] and Stanley[1] urge the application of Mayr's[4,5] paradigm of allopatric speciation to the interpretation of the fossil record. Less emphasis is placed on parapatric and sympatric modes of speciation (cf. refs 6, Ch. 6; 7, 8). According to Mayr, gene flow and genetic homeostasis confer 'genetic cohesion' on populations of a species and prevent significant phenotypic divergence until an isolated population is established in a new environment by a few individuals. The 'founder effect' is supposed to deplete most of the genetic variation and (sometimes) to trigger a 'genetic revolution' entailing rapid and substantial morphological change with reproductive isolation as a by-product. The new species may then expand its range, perhaps supplanting the parent species over part or all of its range.

From recent development and applications of quantitative genetic models, it appears that Mayr's theory of genetic cohesion and the founder effect is largely incorrect. Instead the following picture is emerging. Migration alone cannot enforce phenotypic uniformity across a species range when this distance is much longer than the mean individual dispersal distance. Stabilizing selection is the most powerful factor promoting phenotypic similarity in time and space. A new colony founded from a small number of migrants drawn at random from a large population is expected to contain a large fraction of the total genetic variation, whether this is defined in terms of heterozygosity or heritable genetic variation available for selection. Alternative mechanisms, based on high rates of spontaneous polygenic mutations, are capable of producing rapid divergence in small isolated populations.

Quantitative characters are generally influenced by many genes of individually small effect, in addition to environmental effects [9-11]. Measured rates of spontaneous mutation for typical quantitative characters are in excess of 10^{-2} per gamete per character per generation. This is orders of magnitude higher than conventional mutation rates for single genes with major effects (about 10^{-6} per locus per generation) because most morphological traits are polygenic, and mutations of small effect occur much more frequently than those with large effect. In units of additive genetic variance, spontaneous mutation typically produces each generation about $10^{-3}\sigma_e^2$, where σ_e^2 is the environmental variance of the character (the phenotypic variance expressed in a genetically homogeneous population). Spontaneous mutation can maintain the levels of heritable variation observed in natural populations, even for characters under strong stabilizing selection. The implications of these findings for the theory of allopatric speciation is that even a small isolated population can generate sufficient genetic variation by mutation for a geologically rapid shift into a new adaptive zone, on a time scale of a few hundred to a few thousand generations [12,13].

Abundant evidence from hybridization experiments shows that morphological differences between closed related species and subspecies usually have a polygenic basis (refs 5, p. 543; 14, pp. 143–148; 15, Ch. 8). It is important to realize that quantitative genetics deals with the totality of phenotypic variation from environmental and genetic sources. This includes both regulatory and structural genes, although these categories are not distinct. The popular argument that a regulatory gene affecting relative growth rates of different structures early in development can produce a major discontinuous morphological change is undoubtedly correct, as many such mutations are known[1, 16-19]. However, there are few (if any) genetically well-established cases of morphological macromutations which have been fixed in natural populations of animals. Mutations of large effect are almost always deleterious due either to their main effect or to pleiotropic side effects (e.g., refs 20, 21). In contrast, simultaneous changes in the frequencies of many genes with small pleiotropic effects allow loci with compensatory action to minimize the expression of deleterious traits in evolution. Fisher[6] (pp. 41–44) cogently reasoned that in the process of adaptation to a specific niche the probability of selective advantage of undirected mutations decreases rapidly with increasing magnitude of their effects, being nearly $\frac{1}{2}$ for small mutations and very low for large mutations. When compounded with the much higher mutation rate of genes with small effects, this explains why morphological evolution should generally be polygenic.

The nature of variation in meristic and threshold characters, like number of vertebrae, digits or teeth, is commonly misunderstood. Although the phenotypic variation of such characters is often quasi-continuous or discontinuous, its genetic basis is usually polygenic and the evolution of such characters is amenable to quantitative genetic analysis[21-24]. The often-cited example of the evolution of an extra joint in the maxillary bone of bolyerine snakes (refs 1, pp. 161–162; 18, 25) is just as easily, and likely more correctly, interpreted as a polygenic mechanism of (partial) reduction in the thickness of the middle of the maxilla past a threshold where it is divided in two. The rate of evolution of a polygenic threshold character is much faster when the variants are at intermediate frequency than when they are rare, because selection can not act as effectively on rare variants as on common ones (refs 9, Ch. 18; 26).

Similar confusion exists concerning variation in allometric growth and developmental fields. That morphological differences between related species could be explained as simple changes in a few growth gradients or developmental fields (as attempted by Davis[27], for the giant panda; cited by Stanley[1], pp. 55–56, 138, 157–158), does not imply that only a few genes were involved. On the contrary, evidence exists that natural variation in parameters of allometric growth and developmental fields is usually influenced by multiple genetic factors acting relatively late in development (refs 9, Ch. 5 & 15; 29–31).

Chromosomal rearrangements have also been invoked as possible regulatory macromutations (refs 1, pp. 145–148 & 179; 16, 32, 33, 34, pp. 405–407), as suggested by a large-scale correlation between rates of chromosomal and morphological evolution. There are, nevertheless, many sibling species which differ by chromosomal rearrangements[35] and the direct evidence on newly arising spontaneous and induced rearrangements shows that in most higher organisms inversions, fusions, and translocations rarely produce noticeable morphological effects but often reduce the fertility of structural heterozygotes, so that cytological tests are required to confirm their existence[36,37]. A correlation of rates of chromosomal and morphological evolution would occur if both are accelerated by population subdivision, which increases the rate of random genetic drift between alternative stable states of a population (ref. 11, p. 473). Aneuploidy and polyploidy are important mechanisms of evolution in many plants and some invertebrates but are usually deleterious or lethal in higher vertebrates when euchromatic segments of appreciable size are involved.

RATES AND PATTERNS OF PHENOTYPIC EVOLUTION

Advocates of punctuated equilibrium and macromutation cite as evidence the frequent absence of transitional forms from the fossil record (refs 1, p. 39; 38, 40). This negative information is not convincing, especially in view of the claim that morphological evolution typically occurs in geographically restricted populations, which are unlikely to be fossilized and discovered. Gaps between fossil or living taxa do not imply that the forms evolved rapidly (see refs 39, 41), or that macromutations were involved (see refs 2, pp. 359–376; 42, Ch. 6; 43). Even supergenes controlling major color polymorphisms probably originated by the accumulation of linked and unlinked modifiers (ref. 44, pp. 293–310).

An instructive example to consider is the evolutionary loss of limbs in reptiles. A naive observer having at hand only the fragmentary fossil record of snakes and lizards would be hard pressed to imagine that evolutionarily stable intermediate forms could exist. But the study of living teiid and scincid lizards reveals several independent genera with parallel series of many intermediate forms spanning the entire transition from typically lizardlike species with complete limbs to limbless snakelike forms. Such diverse morphological types are classified in a single genus precisely because there are no obvious gaps at which to split the series. Although these extant genera in the transitional stages of limb loss have no known fossil record, comparison with other genera of reptiles with fossil records suggests an average age of at least several million years. Therefore it is likely that the major morphological change to limblessness is on the whole very slow, involving many genes[26]. Similarly, loss of eyes in a species of cave fish was demonstrated by breeding experiments to involve several genes[45] although it might have occurred rapidly through selective migration towards the light by individuals with the greatest visual acuity (ref. 6, pp. 142–143). If simple reduction and loss of characters results from the accumulation of many genetic factors of small effect, one should also expect the evolution of new complex adaptations to be polygenic.

Proponents of punctuated equilibrium assert that the direction of morphological changes during speciation is random with respect to the long-term trends of evolution which are determined by selection between species (cf. ref. 46, Ch. 4). They utilize an analogy of Wright[47] (p.

121) and Mayr[5] (p. 621) that compares speciation with mutation, and individual selection with species selection. Stanley (refs 1, p. 187; 48) thereby proposes that 'macroevolution is decoupled from microevolution,' while Gould and Eldredge[38] (p. 139) deny that gene frequency changes within populations are the foundation of major evolutionary events. But the principles of population genetics apply both to small isolated populations and to large widespread ones, and even macromutations must obey the rules of inheritance. The pattern of genetic variation within populations is crucial in determining the rate and direction of phenotypic evolution. Multiplication and extinction of lineages do not themselves create complex morphologies, which can only arise by the accumulation of genetic changes in a continuous line of descent that through time may be shifting in space and fluctuating in size.

In contrast with Williams[49], who postulated that nearly all evolution results from individual selection, Stanley, Gould and Eldredge regard long-term trends as due mainly to species selection. More balanced views suggest that when these processes are quantified their relative magnitudes will depend strongly on details of population structure and demography (e.g., refs 50–52). It should not be forgotten that, excepting certain forms of non-Mendelian inheritance, genetic variation between populations arises from genetic variation within populations, and that, excluding phyletic transformation, the extinction of a species coincides with the death of all its members.

Stanley[1] (p. 62, 183–184, 279) attributes the vast majority of phenotypic evolution to selection and assigns little significance to probabilistic factors such as random genetic drift and random extinction. It is possible, however, to examine whether the direction of morphological change within lineages is random by applying standard statistical tests to time-series data on quantitative characters. Van Valen's[53] demonstration of nearly constant rates of extinction within higher taxa may reflect a large element of chance.

There are serious problems with Stanley's methods of estimating taxonomic rates of evolution, and the use of exponential growth models in this context may be oversimplified[54,55]. The methods of Raup[56] and Van Valen[57] also seem inadequate since they apply only to groups with a stable level diversity, $r = 0$, where the steady-state survivorship curve as a function of age x for extinct taxa, l_x, is proportional to the stable age distribution of living taxa.

$$e^{-rx}l_x \Big/ \int_0^\infty e^{-rx}l_x dx.$$

Gould and Eldredge[38] (p. 134) and Stanley[1] (pp. 56–57) interpret the extremely low minimum selective mortalities necessary to explain long-term evolution in the dimensions of mammalian molar teeth[58] as strong evidence against phyletic gradualism. That a few selective deaths per million individuals per generation is sufficient to account for the evolution of molar tooth form during the entire history of horses suggests to me only the tremendous power of selection in determining the course of evolution within a lineage. Stanley[1] (pp. 48–51) feels that spatial variation in selection combined with local gene flow will largely nullify directional selection in large populations. He downplays the possibility of selective pressures that are spatially and temporally sustained in direction, if not also in magnitude, i.e. the conventional explanation of convergent and parallel evolution (ref. 2, Ch. 8).

Body size was singled out by Gould and Eldredge[38] as the character that is most likely to undergo gradual phyletic evolution, presumably because of its general ecological significance, although from a genetic point of view body size is a typical polygenic character. Stanley[1] (pp. 98–99) identifies sexual selection as one factor that would consistently favor large body size throughout a species range. Labelled as a gradualist by Stanley[1] (p. 209), Fisher[6] (pp. 151–153) described a runaway process of sexual selection that could cause a rapid burst of evolution followed by a long period of stasis. This involves unstable positive feedback between female mating preferences and secondary sexual characters of males. The efficacy of this process is supported by observations that females of various insects and vertebrates do exercise mating preferences[59] and that closely related species in these taxa often differ mainly in the secondary sexual characters of males in ways which cannot be fully explained by other selective agencies (ref. 60, Ch. 8).

A variety of mechanisms exist within the framework of established evolutionary theory which can produce a pattern of rapid morphological change, and/or splitting of lineages, followed by prolonged stasis. A changing environment and pattern of selection is the most obvious, e.g., after colonization of a new habitat[2,5]. Selective mating and selective migration may be equally important as modes of rapid speciation without geographic isolation[22]. In small populations, random genetic drift between adaptive zones could leave gaps in the fossil record[2,61]. Discontinuous variation in a polygenic threshold character produces a rate of evolution which is necessarily uneven in time[9].

The relative importance of rapid vs. gradual evolution is basically an empirical question that cannot be resolved by theorizing or speculation. However, the direct evidence from genetic experiments already demonstrates that morphological evolution usually has a polygenic basis. Thus, regardless of the rates of evolution involved, polygenic changes are more plausible than macromutations to account for most morphological diversity in higher animals.

1. Stanley, S.M. *Macroevolution: Pattern and Process.* W. H. Freeman and Co.; San Francisco (1979).

2. Simpson, G. G. *The Major Features of Evolution.* Columbia Univ. Press; New York (1953).

3. Eldredge, N. & Gould, S. J. Punctuated equilibria: an alternative to phyletic gradualism. Pp. 82–115. In: Schopf, T. J. M., ed. *Models in Paleobiology.* Freeman, Cooper and Co.; San Francisco, Calif. (1972).

4. Mayr, E. Change of genetic environment and evolution. Pp. 157–180. In: Huxley, J., Hardy, A. C. & Ford, E. B. eds. *Evolution as a Process.* Allen and Unwin; London (1954).

5. Mayr, E. *Animal Species and Evolution.* Harvard Univ. Press; Cambridge, Mass. (1963).

6. Fisher, R. A. *The Genetical Theory of Natural Selection.* 2nd. ed. Dover; New York (1958).

7. Endler, J. A. *Geographic Variation, Speciation and Clines.* Princeton Univ. Press; Princeton, New Jersey (1977).

8. White, M. J. D. *Modes of Speciation.* Freeman; San Francisco, Calif. (1978).

9. Falconer, D. S. *Introduction to Quantitative Genetics.* Robert MacLehose and Co.; Glasgow (1960).

10. Wright, S. *Evolution and the Genetics of Populations.* Vol. 1. Genetic and Biometric

Foundations. Univ. Chicago Press; Chicago, Ill. (1968).

11. Wright, S. *Evolution and the Genetics of Populations.* Vol. 3. Experimental Results and Evolutionary Deductions. Univ. Chicago Press; Chicago. Ill. (1977).

12. Lande, R. The maintenance of genetic variability by mutation in a polygenic character with linked loci. *Genet. Res.* **26**, 221–235 (1976).

13. Lande, R. Genetic variation and phenotypic evolution during allopatric speciation. *Am. Nat.* (In press) (1980).

14. Dobzhansky, Th. *Genetics and the Origin of Species.* 3rd ed. Columbia Univ. Press; New York (1951).

15. Wright, S. *Evolution and the Genetics of Populations.* Vol. 4. Variability within and among Natural Populations. Univ. Chicago Press; Chicgo. Ill. (1978).

16. Goldschmidt, R. *The Material Basis of Evolution.* Yale Univ. Press; New Haven, Conn. (1940).

17. Valentine, J. W. & Campbell, C. A. Genetic regulation and the fossil record. *Am. Sci.* **63**, 673–680 (1975).

18. Gould, S. J. The return of hopeful monsters. *Nat. Hist.* **86**, 22–30 (1977).

19. Alberch, P., Gould, S. J., Oster, G. F. & Wake, D. B. Size and shape in ontogeny and phylogeny. *Paleobiology.* **5**, 296–317 (1979).

20. Caspari, E. Pleiotropic gene action. *Evolution.* **6**, 1–18 (1952).

21. Grüneberg, H. *The Pathology of Development.* Blackwell Sci. Publ.; Oxford (1963).

22. Fisher, R. A. *The Genetical Theory of Natural Selection.* 1st ed. Oxford Univ. Press; Oxford (1930).

23. Wright, S. The results of crosses between inbred strains of guinea pigs, differing in number of digits. *Genetics.* **19**, 537–551 (1934).

24. Green, E. L. Quantitative genetics of skeletal variations in the mouse. II. Crosses between four inbred strains. *Genetics.* **47**, 1085–1096 (1962).

25. Frazetta, T. H. From hopeful monster to bolyerine snakes? *Am. Nat.* **104**, 55–72 (1970).

26. Lande, R. Evolutionary mechanisms of limb loss in tetrapods. *Evolution.* **32**, 73–92 (1978).

27. Davis, D. D. The giant panda: a morphological study of evolutionary mechanisms. *Fieldiana Mem. (Zool.)* **3**, 1–339 (1964).

28. Kidwell, J. F., Gregory, P. W. & Guilbert. H. R. A genetic investigation of allometric growth in Hereford cattle. *Genetics.* **37**, 158–173 (1952).

29. Kidwell, J. F. & Williams, E. Allometric growth of the Dark Cornish fowl. *Growth.* **20**, 275–293 (1956).

30. Cock, A. G. Genetical aspects of metrical growth and form in animals. *Q. Rev. Biol.* **41**, 131–190 (1966).

31. Cock, A. G. Genetical studies on growth and form in the fowl. *Genet. Res.* **14**, 237–247 (1969).

32. Wilson, A. C., Sarich, V. M. & Maxson, L. R. The importance of gene rearrangement in evolution: evidence from studies on rates

of chromosomal, protein, and anatomical evolution. *Proc. Natl. Acad. Sci. USA.* **71**, 3028–3030 (1974).

33. Bush, G., Case, S. M., Wilson, A. C. & Patton, J. L. Rapid speciation and chromosomal evolution in mammals. *Proc. Natl. Acad. Sci. USA* **74**, 3942–3946 (1977).

34. Gould, S. J. *Ontogeny and Phylogeny.* Harvard Univ. Press; Cambridge, Mass. (1977).

35. White, M. J. D. *Animal Cytology and Evolution.* Cambridge Univ. Press; Cambridge (1973).

36. Lande, R. Quantitative genetic analysis of multivariate evolution, applied to brain: body size allometry. *Evolution.* **33**, 402–416 (1979).

37. Lande, R. Effective deme sizes during longterm evolution estimated from rates of chromosomal rearrangement. *Evolution.* **33**, 234–251 (1979).

38. Gould, S. J. & Eldredge, N. Punctuated equilibria: the tempo and mode of evolution reconsidered. *Paleobiology.* **3**, 115–151 (1977).

39. Gingerich, P. D. Patterns of evolution in the mammalian fossil record. Pp. 469–500. In: Hallam, A., ed. *Patterns of Evolution.* As Illustrated by the Fossil Record. Elsevier; Amsterdam (1977).

40. Gish, D. *Evolution. The Fossils Say No!* Creation-Life Publ.; San Diego, Calif. (1973).

41. Gingerich, P. D. Evolutionary transition from ammonite *Subprionocyclus* to *Reesidites*—punctuated or gradual? *Evolution.* **32**, 454–456 (1978).

42. Darwin, Ch. *On the Origin of Species.* [Facsimile of the 1st ed.] Harvard Univ. Press; Cambridge, Mass. (1859).

43. Mayr, E. The emergence of evolutionary novelties. Pp. 349–380. In: Tax, S., ed. *Evolution after Darwin.* Vol. 1. Univ. Chicago Press; Chicago. Ill. (1959).

44. Ford, E. B. *Ecological Genetics.* 4th ed. Wiley; New York (1975).

45. Wilkins, H. Genetic interpretation of regressive evolutionary processes: studies on hybrid eyes of two *Astyanax* cave populations (Characidae, Pices). *Evolution.* **25**, 530–544 (1971).

46. Rensch, B. *Evolution above the Species Level.* Columbia Univ. Press; New York (1959).

47. Wright, S. Comments on the preliminary working papers of Eden and Waddington, Pp. 117–120 In: Moorhead, P. S. & Kaplan, M. M. eds. *Mathematical Challenges to the Neo-Darwinian Interpretation of Evolution.* Wistar Inst. Press; Philadelphia, Pa. (1967).

48. Stanley, S. M. A theory of evolution above the species level. *Proc. Natl. Acad. Sci. USA.* **72**, 646–650 (1975).

49. Williams, G. C. *Adaptation and Natural Selection.* Princeton Univ. Press; Princeton, New Jersey (1966).

50. Lewontin, R. C. The units of selection. *Annu. Rev. Ecol. Syst.* **1**, 1–18 (1970).

51. Wade, M. J. A critical review of the models of group selection. *Q. Rev. Biol.* **53**, 101–114 (1978).

52. Leigh, E. G., Jr. How does selection reconcile individual advantage with the good of the group? *Proc. Natl. Acad. Sci. USA.* **74**, 4542–4546 (1978).

53. Van Valen, L. A new evolutionary law. *Evol. Theory.* **1**, 1–30 (1973).
54. Gillespie, J. H. & Ricklefs, R. E. A note on the estimation of species duration distributions. *Paleobiology.* **5**, 60–62 (1979).
55. Ricklefs, R. E. Paleontologists confronting macroevolution. *Science.* **199**, 58–60 (1978).
56. Raup, D. M. Taxonomic survivorship curves and Van Valen's law. *Paleobiology.* **1**, 82–96 (1975).
57. Van Valen, L. Taxonomic survivorship curves. *Evol. Theory.* **4**, 129–142 (1979).
58. Lande, R. Natural selection and random genetic drift in phenotypic evolution. *Evolution.* **30**, 314–334 (1976).
59. Ehrman, L. Genetics and sexual selection. Pp. 105–135. In: Campbell, B., ed. *Sexual Selection and the Descent of Man.* Aldine; Chicago, Ill. (1972).
60. Darwin, Ch. *The Descent of Man, and Selection in relation to Sex.* 2nd. ed. John Murray; London (1874).
61. Wright, S. Breeding structure of populations in relation to speciation. *Am. Nat.* **74**, 232–248 (1940).

Russell Lande is in the Department of Biophysics and Theoretical Biology, The University of Chicago.

This article first appeared in *Paleobiology* Vol. **6**, pp. 235–238; 1980. It is reproduced with permission.

Palaeontological documentation of speciation in Cenozoic molluscs from Turkana Basin

P. G. WILLIAMSON

A recently discovered series of mollusc faunas from the late Cenozoic of the eastern Turkana Basin constitutes one of the best documented metazoan fossil sequences. Evolutionary patterns in all lineages conform to the 'punctuated equilibrium' model; no 'gradualistic' morphological trends occur. These faunas provide the first fine-scaled palaeontological resolution of events during speciation: fundamental phenotypic transformation of both sexual and asexual taxa occurs rapidly, in comparatively large populations, and is accompanied by a significant elevation of phenotypic variance. This increase in variance reflects extreme developmental instability in the transitional populations.

THE 400 m sequence of late Cenozoic deposits east of Lake Turkana in northern Kenya, discovered[1] by Leakey in 1968, comprises the Plio-Pleistocene Kubi Algi, Koobi Fora and Guomde Formations, and the Holocene Galana Boi Beds[2]. In addition to important palaeoanthropological[1], archaeological[3] and vertebrate palaeontological[4] material, these deposits have recently yielded a uniquely well documented sequence of lacustrine mollusc faunas. The faunas have important implications for present evolutionary controversies, as they provide the first detailed palaeontological documentation of events during allopatric speciation.

Mollusc faunas are scattered throughout 1,000 km² of exposures east of Lake Turkana and occur in laterally extensive lensoid accumulations 0.01–1 m thick. These accumulations have a matrix ranging from coarse silt to coarse sand grade, and are separated by finer-grained intervals devoid of molluscs. The 190 faunas reported here consist of various prosobranch, pulmonate and bivalve lineages, and represent both life and death assemblages in various shallow lacustrine and pro-deltaic settings.

Various features make this sequence particularly useful for investigating evolutionary patterns: the molluscs are well preserved and abundant, the units in which they occur are generally unconsolidated, and because most of the species lineages in the section are still extant, reasonable inferences can be drawn concerning the 'soft' biology of their fossil representatives from the Turkana Basin sequence. In biological and taxonomic terms, these mollusc lineages comprise an extremely heterogeneous assemblage (see Table 1); this heterogeneity allows useful comparisons to be made of evolutionary patterns in taxa varying widely in autecology, reproductive strategy and size. In particular, evolutionary patterns in both sexual and asexual taxa can be compared. The 19 species lineages in the section represent 18 genera and 12 families, thus ancestor-descendant relationships between species lineages and their derivative taxa are unambiguous. Because the molluscan shell accretes terminally and is normally unmodifiable after deposition, each shell is a comprehensive record of individual development, and it is therefore unneccessary to construct mass curves to study changes in patterns of ontogeny during evolution. A useful general scheme for the quantification of shell form is available[5]. Finally, previous extensive geological investigations in the Turkana Basin mean that the mollusc sequence can be studied within a well documented chronostratigraphical and palaeoenvironmental context. A comparative evolutionary study can be made within a quasi-experimental context.

COLLECTION AND TREATMENT OF MATERIAL

Of the 19 species lineages, 13 were common enough for a biometric analysis. 2–10 kg of shell-bearing sediment were extracted from each fauna, from which I selected random subsets of individuals representing the various species lineages. 20–30 individuals of each lineage in the sample were usually measured but shell breakage sometimes reduced this number.

Images of selected shells were projected onto a System-2 D-MAC digitizing table and 5–24 (average 18) measurements, depending on the lineage studied, were made for each shell. These measurements included the cartesian coordinates of specific points on the shell, and the areas, perimeters and geometric centroids of the generating curve and muscle scars. 5–20 (average 15) parameters were derived for each individual from the original measurements, and formed the basis of the biometric analysis and included approximations to Raup's parameters[5] W (whorl expansion rate), T (translation rate) and D (distance of generating curve from coiling axis), all taken at 2-rad intervals down the coiling axis of each shell. Aspects of Raup's parameter S (shape of generating curve) and details of shell sculpture were also recorded; for the bivalves, the derived parameters also included descriptions of shape, relative size and disposition of the muscle scars and hinge features. Some 3,300 individuals representing 13 lineages were measured. Canonical variate and principal component analyses were performed on the derived parameters for each species lineage.

The Turkana Basin sequence is punctuated by a prominent series of tuffaceous horizons. The bases of these tuffs are isochronous, laterally extensive and mappable[6], and they form the basis of the regional chrono-

Table 1 Primary biological properties of mollusc species lineages from the Turkana Basin sequence

Species	Group	Infaunal	Epifaunal	Deposit feeder	Suspension feeder	Gonochoristic (internal fertilization)	Gonochoristic (external fertilization)	Facultative hermaphrodite	Permanently thelytokous	Modal adult length (mm)
Bellamya unicolor*	Prosobranchs		×	×		×				30
Pila ovata	Prosobranchs		×	×		×				65
Valvata sp.*	Prosobranchs		×	×				×		4
Gabiella senaariensis*	Prosobranchs		×	×		×				2.5
Cleopatra ferruginea*	Prosobranchs		×	×		×				17
Melanoides tuberculata*	Prosobranchs	×		×					×	15
Bulinus truncatus*	Pulmonates		×	×				×		8
Gyraulus costulatus	Pulmonates		×	×		×				55
Burnupia sp.	Pulmonates		×	×		×				3
Caelatura (Caelatura) bakeri*	Bivalves	×			×		×			35
Caelatura (nitia) monceti*	Bivalves	×			×		×			30
Pseudobovaria sp.*	Bivalves	×			×		×			16
Aspatharia arcuta	Bivalves	×			×		×			160
Mutela nilotica*	Bivalves	×			×		×			110
Pleiodon sp.*	Bivalves	×			×		×			100
Etheria elliptica	Bivalves		×		×		×			140
Corbicula consobrina*	Bivalves	×			×		×			14
Pisidium (afropisidium) piroth	Bivalves	×			×		×			2
Eupera ferruginea*	Bivalves		×		×		×			7

Data are from various sources.
*Taxa used in the biometric analyses.

stratigraphic framework[2]. Mollusc faunas were therefore stratigraphically ordered by surveying them into tuff bases. This ordering of the
faunas assumes that sedimentation rates in the contiguous areas from
which they were collected were similar—apparently the case—and that
the various tuffs have been correctly correlated throughout the areas of
exposure. Some correlations between tuffs have been questioned on the
basis of (vertebrate) biostratigraphical evidence[7,8], but these problematic
correlations are not generally in areas yielding the molluscan faunas.

PATTERNS OF EVOLUTIONARY CHANGE

Despite the variation in autecology, reproductive strategy and size of the
lineages, their patterns of evolutionary change are fundamentally identical. Figures 2 and 3 summarize aspects of this pattern for the prosobranch
Bellamya unicolor (Olivier). Canonical variate analysis indicates that populations from the Kubi Algi and pre-Kubi Algi, from most levels in the
Koobi Fora Formation and from the Galana Boi Beds group together (Fig.
2). All individuals from these populations are readily referred to the
extant species *B. unicolor*. Despite the long-term stasis shown by the *B.
unicolor* lineage during the later Cenozoic, three stratigraphically circumscribed but important morphological excursions occur during this period:
at the Suregei tuff level, in the Lower Member of the Koobi Fora Formation and in the Guomde. At the Suregei tuff (Fig. 1), populations of
Bellamya from faunas 12b and 12c consist exclusively of a novel and highly
distinct form. Populations morphometrically intermediate between this
unusual form and typical *B. unicolor* occur in faunas 8 and 11, just below
the base of the Suregei tuff. Another morphological excursion occurs in
faunas 27 and 29 in the Lower Member of the Koobi Fora Formation; here
a second novel form of *Bellamya* occurs together with typical *B. unicolor*. A
third excursion is documented by fauna 88 in the Guomde, where populations consist exclusively of a third novel form of *Bellamya*. Although Fig.
2 suggests that the Guomde population resembles that of the aberrant
Suregei tuff level, it is in fact displaced away from them along the third
canonical variate; in terms of generalized distance (Mahalanobis' D^2), the
Guomde population is equidistant from the latest Suregei level *Bellamya*
(6.25 units from population 12c) and the nearest typical *B. unicolor* population (6.35 units from population 30). Intermediates between typical *B.
unicolor* and the divergent *Bellamya* morphs in faunas 27 and 29, and in
fauna 88, are unknown.

As suggested by Fig. 2 and confirmed by comparisons with museum
collections of recent *B. unicolor* material, all three novel morphs of
Bellamya lie outside the narrow phenotypic range of *B. unicolor*. The
profound differences in shell geometry of these divergent forms are at
least as great as those characteristic of different extant *Bellamya* species.
However, the presence of the intermediate forms at the Suregei tuff level
and the absence of other potentially ancestral forms of *Bellamya* in the late
Cenozoic of north-east Africa indicate that all three divergent morphs of
Bellamya in the section are derived from *B. unicolor*, although the details of
this derivation are only documented at the Suregei tuff level.

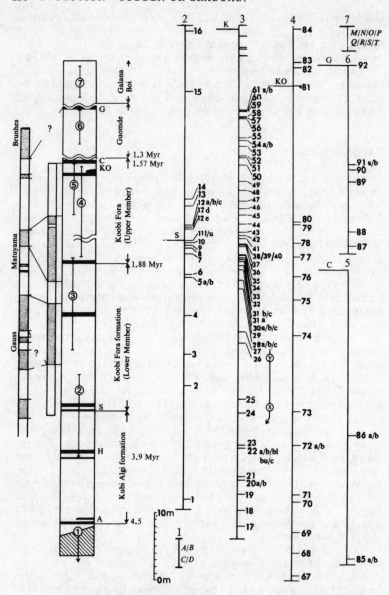

FIG. 1. *Generalized chronostratigraphy of late Cenozoic deposits of the eastern Turkana Basin and stratigraphical distribution of mollusc faunas. Geophysical age data and generalized stratigraphy are from various sources*[2,12,38-42]. *On the extreme left is shown the world geomagnetic polarity time scale; to the right of this is the polarity sequence observed in the eastern Turkana Basin. Generalized stratigraphical column shows principal tuff horizons and (to the right)* [39]Ar/[40]Ar or K/Ar

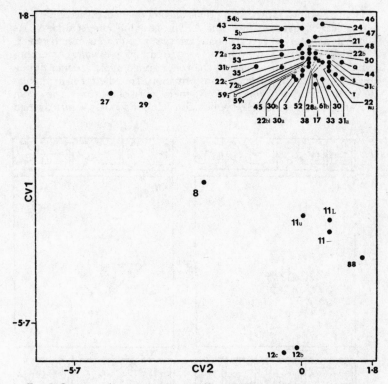

FIG. 2. *Summary of canonical variate analysis of 49 populations of the* B. unicolor *species lineage from the late Cenozoic of the Turkana Basin. Analysis based on 16 parameters derived from 16 original measurements, for each of 761 individual specimens. Population centroids are plotted on to the first two canonical variates (CV1 and CV2), which together explain ~ 60% of the total variance. Population numbers correspond to fauna numbers in Fig. 1 (See discussion in text.)*

age determinations of the tuffs. A, Alia tuff; H, Hasuma tuff; S, Suregei tuff; G, top Guomde tuff. Numbered subsections on the generalized stratigraphical column indicate the general position of the expanded sections numbered 1–7 in the right half of the diagram. These latter sections show the stratigraphical order of mollusc faunas; the metre scale refers to sections 2–6. Faunas A–D are from deposits thought to pre-date the Kubi Algi, or to correspond to a lower level of that unit (fauna A, Casa Waterhole section[43]; fauna D, deposits at Sibiloit[10]; faunas C, D, Warata Formation[44]). The relative stratigraphical order of these faunas is un-known, as is the order of faunas M–T from the relatively thin Galana Boi Beds. The stratigraphical order of faunas 1–92 is determined by the distance of their upper surfaces from the bases of the named tuff units, as measured in the various local sections. Faunas X and Y come from low down in the Lower Member, but their precise position is unclear. Most of the faunal units are < 0.25 m thick. a, b, c and so on denote faunas that are demonstrably sampled from the same mollusc-bearing unit, but from different localities; earlier letters in the alphabet denote more southerly locations. l and u denote samples from the basal and upper thirds, respectively, of a given faunal unit.

Figure 3 shows aspects of a principal component analysis of the morphological excursion in the *B. unicolor* lineage at the Suregei tuff level. As suggested by the canonical variate analysis, populations from faunas 8, 11 and 12 show rapid movement away from typical *B. unicolor* morphology. There are also significant changes in phenotypic variance during this period of morphological transformation. The intermediate population 8, the centroid of which is 6.66 generalized distance units away from the latest divergent *Bellamya* population (12c), shows substantial overlap

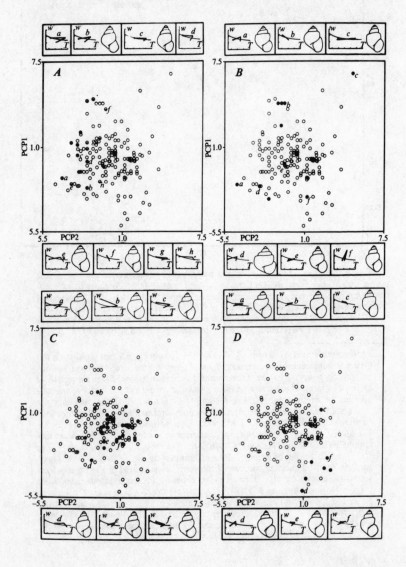

with both typical *B. unicolor* and the latest derivative *Bellamya* morphs (Fig. 3). However, population 8 displays a striking elevation of phenotypic variance (the summed variance of all measured parameters). Total phenotypic variance in population 8 is 45.5, which is significantly greater at the 2.5% level (one-sided *F*-test)[10] than that of populations of typical *B. unicolor* from elsewhere in the section (average variance 9.1, range 5.4–12.3). As discussed below, increase in phenotypic variance reflects major developmental instability in this transitional population. Population 11, 4.39 generalized distance units away from the latest derivative *Bellamya* of fauna 12c, shows greater overlap with this population than fauna 8, but less overlap with typical *B. unicolor*. However, population 11 has a phenotypic variance (12.7) comparable with that of typical *B. unicolor*, despite its intermediate morphology. The latest divergent population from the Suregei tuff level, from fauna 12c, is ~6.9 generalized distance units from the nearest population of typical *B. unicolor* (population 30) and forms a completely non-overlapping cluster with typical *B. unicolor*; however, the variance (18.5) of fauna 12c is comparable with that of typical *B. unicolor*. The general picture of events in the *Bellamya* lineage at the Suregei tuff level is one of rapid morphological transformation, initially accompanied by a major increase in phenotypic variance. However, faunas overlying this period of transformation (fauna 13 and above) yield populations of typical *B. unicolor*.

The general pattern of phenotypic change exhibited by the *B. unicolor* lineage is summarized in Fig. 4, which also indicates similar patterns for all other well documented mollusc lineages. Major phenotypic transformations occur simultaneously in all lineages at the Suregei tuff level. For *B. unicolor*, *Melanoides tuberculata* and *Caelatura bakeri* this change involves a significant initial increase in phenotypic variance (in fauna 8 in *Bellamya* and *Caelatura*, and in faunas 8, 9 and 11 in *Melanoides*). Subsequent transformation is accompanied by reduced levels of phenotypic variance, comparable with those of typical ancestral populations of the

---

FIG. 3. *Summary of principal components analysis of individuals of the* B. unicolor *species lineage from populations at and below the Suregei tuff level, represented by circles, plotted on to the first two principal components (PCP1 and PCP2), which together explain ~ 45% of the total variance.* ● *and* ◒ *represent individuals from five successive faunas spanning the marked phenotypic transformation documented at the Suregei tuff level. A, Individuals from fauna B (●) and from fauna 3 (◒); B, individuals from fauna 8 (●); C individuals from fauna 11 (●); D, individuals from fauna 12c (◓). Note the rapid shift in phenotype from the ancestral* B. unicolor *morphology (in the left-hand half of the plot) to the divergent form from fauna 12c (to the right). Note also the initial increase in phenotypic variance during this transformation, reflected in the wide scatter of individual morphologies from population 8b. The bivariate diagrams above and below the plots show successive ontogenetic values of* W *(whorl expansion rate) and* T *(translation rate)[5] for selected representative individuals. The curve shown is the function* $T = 2\sqrt{W}/(W - 1)$. *The solid point at the end of each W/T track indicates the earliest measured point in ontogeny. These tracks record W and T values at three successive, equivalent,* 2π *radian intervals down the shell (over the last four whorls). Individuals to which the W/T plots and sketches refer are identified on the PCP plots by the letters a, b, and so on. Note the derangement of the W/T tracks in the phenotypically variable intermediate population 8b. (See discussion in text.)*

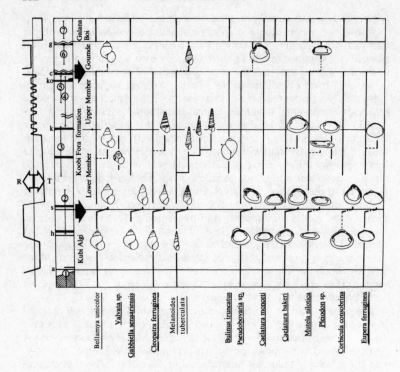

FIG. 4. *Generalized summary of patterns of evolutionary change in the Turkana Basin mollusc sequence. On the left is a generalized stratigraphical section as in Fig. 1. The heavy line on the extreme left of the diagram is a generalized "transgression (T)-regression (R)" line based on various sources[2,11-14]. The lighter vertical line corresponds to a lake level which is largely the same as the present level. Heavy vertical lines in the main part of the diagram indicate the time ranges of the various species lineages. Broken heavy lines at the Suregei tuff level in some lineages indicate poorly documented speciation events. Only the commonest morphologies developed during the extensive sympatric radiation of the asexual form Melanoides tuberculata in the Lower Member are indicated. Some species lineages are not known from the lower- and mid-Kubi Algi Formation, but are known from either pre-Kubi Algi sections in the eastern Turkana Basin, or from the Mursi Formation in the Omo sequence to the north of the basin (D. Vandamme, personal communication). Biostratigraphical evidence (P.G.W., unpublished results) indicates that the latter sequence overlaps with the lower- and mid-Kubi Algi. Although Cleopatra ferruginea is unknown from the Galana Boi, this species is extant in other areas of East Africa. Note that sketches of representative shells are not all drawn to the same scale. (Species lineages whose trivial names are not given (sp.) are now being described by Dr D. Vandamme.) The principal evolutionary events documented in the section are (1) the simultaneus speciation events in all lineages at the Suregei tuff and Guomde levels of the basin (indicated by heavy arrows) and (2) the adaptive radiation of several stocks in the middle part of the sequence (Lower Member of the Koobi Fora Formation).*

species lineage. In *Gabbiella senaariensis* and *Cleopatra ferruginea* the peak of phenotypic variance does not occur until the latest divergent faunas (faunas 12a–d) and its subsequent decline is therefore not documented. A major phenotypic transformation is documented for *Psuedobovaria* sp. A., *Caelatura monceti*, *Corbicula consobrina*, *Mutela nilotica* and *Eupera ferruginea*, but populations intermediate between typical representatives of these species lineages, and their distinctive derivatives at the Suregei tuff level, are too poorly documented for analysis of change in phenotypic variance. By fauna 12, all lineages present have transformed such that, in terms of individual morphology, they constitute completely non-overlapping clusters with the typical ancestral forms: the fauna as a whole is unique and endemic to the Turkana Basin. A similar series of events occurs in the Guomde Formation: all representatives of the various species lineages are endemic and distinct, but the angular unconformity between the uppermost Koobi Fora and the Guomde precludes documentation of intermediate forms.

As indicated in Fig. 4, immediately above the Suregei tuff level there is an abrupt reversion to ancestral morphology in all species lineages. No intermediates are known. Subsequently, in the Lower Member of the Koobi Fora, there is a minor adaptive radiation in several lineages. Novel endemic forms suddenly enter the record and coexist with their parent forms. However, in *Melanoides*, various novel morphs arise sympatrically, via intermediate forms. This phenomenon, described in detail elsewhere[11], apparently represents a gradual clonal radiation in this asexual form. All novel Lower Member lineages are terminated by climatically induced regression[11,12] and the consequent increase in alkalinity[13,14] in the uppermost Lower Member. Intralacustrine endemic radiations such as that documented in the Lower Member of the Koobi Fora are known from many modern rift lakes[15,16].

IMPLICATIONS

A persistent problem in evolutionary biology has been the absence of intermediate forms in the fossil record. Long-term gradual transformations of single lineages are rare[17] and generally involve simple size increase or trivial phenotypic effects[17-20]. Typically, the record consists of successive ancestor–descendant lineages, morphologically invariant through time and unconnected by intermediates. Eldredge and Gould[17] have suggested that this 'punctuated equilibrium' geometry of phylogeny is a logical outcome of several postulated allopatric speciation mechanisms, particularly Mayr's[21] 'founder effect' speciation model. This model considers that homeostatic mechanisms and gene flow prohibit significant evolutionary change in large panmictic species populations, but in small, stressed, geographically isolated populations, homeostatic mechanisms break down during 'genetic revolution' and rapid evolution may ensue[21]. The abrupt entry of new lineages into the fossil record therefore represents immigration from isolated sites of origin; the small size and ephemerality of the populations in which cladogenesis occurs normally preclude palaeontological documentation of speciation and associated intermediate forms[17]. The phylogenetic geometry of molluscan lineages from the Turkana Basin sequence clearly conforms to the punctuated equilibrium model: long-term stasis in all lineages is punctuated by rapid episodes of major phenotypic change. No 'gradualistic' morphological trends[17] occur in any lineage. The long-term temporal stasis in phenotype documented in both sexual and asexual lineages is paralleled by the geographical stability in phenotype exhibited by their widely distributed modern representatives; shell form in the lineages studied is normally highly 'heritable' (the extreme ecophenotypic variability of freshwater Bassomatophora is not typical of the gastropods studied here, most of which are prosobranch species characterized by narrow phenotypic ranges in modern faunas).

Events during the radiation of several lineages in the Lower Member of the Koobi Fora are typical of the fossil record in that intermediates between derivative and ancestral taxa are not documented. However, events at the Suregei tuff and Guomde levels are significant in that they provide, for the first time, details of allopatric speciation during the 'punctuation' of cladogenesis; they allow an unprecedented resolution of the fine structure of events during speciation. Previous documentation of phenotypic divergence within lineages has involved sympatric stocks and has been interpreted as long-term character displacement after unrecorded allopatric speciation events (for example, see refs 22, 23). Although Ovcharenko[24] has claimed to document allopatric derivation of the brachiopod *Kutchithyris euryptycha* from *K. acutiplicata* in a Bathonian section in the Pamirs, no biometric data are available for this supposed transition and the significance of these faunas is unclear.

The events of the Suregei tuff and Guomde levels clearly document speciation within peripheral isolates. Major phenotypic transformations, of at least as great a magnitude as those now characterizing different extant biological species of the genera concerned, occur almost simultaneously in all lineages at these levels, but typical, unaltered representatives of the various ancestral lineages are known from other, contemporaneous East African sites throughout the late Cenozoic[10]. The phenotypic shifts at the Suregei tuff level are continuous, unidirectional and must (from general section thickness estimates) have taken $5 \times 10^3 - 5 \times 10^4$ years to accomplish[10]. I believe that these considerations, together with the long-term temporal stasis in morphology exhibited by these lineages and their current morphological stability in a diverse range of modern environments, preclude the possibility that such dramatic phenotypic shifts are in any sense 'non-genetic' or ecophenotypic.

Although several aspects of Mayr's 'founder effect' model[21] of speciation have been challenged recently[25-27], the pattern of phenotypic change at the Suregei tuff level—rapid directional change in morphology initially accompanied by extreme developmental instability—agrees with important aspects of Mayr's model. As indicated above, Mayr[21] emphasized both geographical isolation and environmental stress as important 'triggers' for speciation. Speciation events at both the Suregei and Guomde levels coincide with major lacustrine regressions, when faunas in the basin are likely to have been both isolated and under stress. Stunting of faunas immediately before the Suregei level and the small number of lineages in the Guomde may reflect stress at these levels. Lacustrine transgression after these regressive phases coincides with reinvasion of ancestral stocks into the basin and elimination of the derivative taxa (Fig. 4): Mayr considers this pattern of events to be the common fate of new species in geographical isolates[21]. In addition, increase in phenotypic variance during allopatric speciation, as documented at the Suregei tuff level, has been predicted as a consequence of Mayr's founder effect model by Levins[28], who suggests that developmental instability, due to breakdown in canalization of individual ontogeny, is a likely concomitant of disruption of homeostatic mechanisms during the 'genetic revolution' accompanying speciation (see also refs 26, 27, 29). The molluscan shell is a terminally accreting structure, and the increase in phenotypic variance documented at the Suregei tuff level therefore reflects disruption of patterns of individual ontogenetic development (this is graphically illustrated by plots of Raup's parameters[5] W and T during ontogenesis—see Fig. 3). However, two major aspects of events at the Suregei tuff and Guomde levels are clearly at variance with Mayr's model: (1) the (obligatory) asexual taxon $M.$ $tuberculata$[30-33] shows a pattern of evolutionary change identical with other sexual species, at both the Suregei tuff and Guomde levels; and (2) evolutionary change at the Suregei level, although occurring rapidly, occurs over a large area and in thick faunal units containing many millions of individuals.

Both these observations question the significance of genetic drift, founder effect and inbreeding as mechanisms for triggering breakdown of homeostasis and speciation in geographical isolates. Moreover, the events in the asexual taxon $Melanoides$ question the conventional assumption that the significance of geographical isolation to speciation is blockage of gene flow (gene flow may in any case be an insignificant phenomenon, even in sexual species)[34]. I propose elsewhere[10] that geographical isolation is a prerequisite for speciation in that it shields transitional populations, vulnerable due to developmental instability, from competition with their unaltered ancestral taxa. The power of density-dependent stabilizing selection (the competitive vulnerability of phenodeviants to more 'modal' phenotypes), well documented by laboratory work[35] and by observations of natural populations[36], is relevant here.

Apart from the tantalizing insights into speciation mechanisms offered by the Turkana Basin sequence, it has two more general implications for evolutionary theory. The documented restriction of significant evolutionary change to speciation events indicates that the underlying unit of macro-evolutionary change is the species. The fact that evolutionary change at the species level is shown to be punctuated and achieved by 'revolutionary' periods of extreme developmental instability strongly supports the notion that speciation is a qualitatively different phenomenon from gradual, intraspecific microevolutionary change[29,37].

This work was carried out at the University of Bristol and was supported by a NERC studentship to P.G.W. and a NERC research grant to Dr R. J. G. Savage. I thank Dr R. J. G. Savage, Miss A. P. Gotto, Mr R. E. F. Leakey, Mrs Suzanne Evans, Mr I. J. Metcalfe and all members of the Koobi Fora research project. Professors J. Maynard Smith, S. J. Gould and E. Mayr provided stimulating discussion. The figures were drawn in part by Mrs J. Bees and Mrs A. Gregory.

1. Leakey, R. E. F. & Leakey, M. (eds) *Koobi Fora Research Project Monographs* 1 (Clarendon, Oxford, 1978).
2. Vondra, C. F. & Bowen, B. E. in *Geological Background to Fossil Man* (ed. Bishop, W. W.) (Scottish Academic, Edinburgh, 1978).
3. Isaac, G. Ll. in *Earliest Man and Environments in the Lake Rudolf Basin* (eds Coppens, Y., Clark Howell, F., Isaac, G. Ll. & Leakey, R. E. F.) (University of Chicago Press, 1976).
4. Harris, J. M. in *Koobi Fora Research Project Monographs* 1 (eds Leakey, R. E. F. & Leakey, M.) (Clarendon, Oxford, 1978).
5. Raup, D. M. *J. Paleont*, **40**, 1178–1190 (1966).
6. Findlater, I. C. in *Geological Background to Fossil Man* (ed. Bishop, W. W.) (Scottish Academic, Edinburgh, 1978).
7. White, T. D. & Harris, J. M. *Science* **198**, 13–21 (1977).
8. Harris, J. M. & White, T. D. *Trans. Am. phil. Soc.* **69**, 1–128 (1979).
9. Simpson, G. G., Roe, A. & Lewontin, R. C. *Quantitative Zoology* (Academic, New York, 1960).
10. Williamson, P. G. thesis, Univ. Bristol (1980).
11. Johnson, G. D. & Raynolds, R. G. H. in *Earliest Man and Environments in the Lake Rudolf Basin* (eds Copens, Y., Clark Howell, F., Isaac, G. Ll. & Leakey, R. E. F.) (University of Chicago Press, 1976).
12. Findlater, I. C. in *Koobi Fora Research Project Monographs* 1 (eds Leakey, R. E. F. & Leakey, M.) (Clarendon, Oxford, 1978).
13. Abell, P. I., Awramik, S. M., Osborne, R. H. & Tomellini, S. *Sediment Geol*. (in the press).
14. Cerling, T. E. *Palaeogeogr. Palaeoclimatol. Palaeoecol*. **27**, 247–285 (1979).
15. Brooks, J. L. *Q. Rev. Biol*. **25**, 132–176 (1950).
16. Gould, S. J. & Eldredge, N. *Paleobiology* **3**, 115–151 (1977).
17. Eldredge, N. & Gould, S. J. in *Models in Paleobiology* (ed. Schopf, T. J. M.) 82–115 (Freeman, Cooper, San Francisco, 1972).
18. Hallam, A. *Nature* **258**, 439–446 (1975).
19. Gould, S. J. & Eldredge, N. *Paleobiology* **3**, 115–151 (1977).
20. Stanley, S. M. *Macroevolution* (Freeman, San Francisco, 1979).
21. Mayr, E. in *Evolution as a Process* (eds Huxley, J., Hardy, A. C. & Ford, E. B.) 157–180 (1954).
22. Gingerich, P. D. *Am. J. Sci*. **276**, 1–28 (1976).
23. Kellog, D. E. *Paleobiology* **1**, 150–160 (1975).
24. Ovcharenko, V. N. *Palaeont, J*. **57**, 57–63 (1969).
25. Lande, R. *Am. Nat*. **116**, 463–479 (1980).
26. Templeton, A. R. *Evolution* **34**, 719–729 (1980).
27. Templeton, A. R. *Genetics* **94**, 1011–1020 (1980).
28. Levins, D. A. *Am. Nat*. **104**, 343–353 (1970).
29. Carson, M. L. *Am. Nat*. **109**, 83–92 (1975).
30. Jacob, J. *Trans. R. Soc. Edinb*. **63**, 341–352 (1957).
31. Jacob, J. *Trans. R. Soc. Edinb*. **64**, 433–444 (1958).
32. Starmühlner, F. *Malacologia* **8**, 1–434 (1969).
33. Mandahl-Barth, G. *Field Guide to African Freshwater Snails*, 2 (WHO Snail Identification Centre, Danish Bilharziasis Laboratory, 1973).
34. Enrlich, P. R. & Raven, P. H. *Science* **165**, 1228–1232 (1969).
35. Barnes, C. & Kearsey, A. L. *Heredity* **25**, 1–21 (1970).
36. Walker, B. W. *Calif. Fish Game* **113**, 104–151 (1961).
37. Gould, S. J. *Paleobiology* **6**, 119–130 (1980).
38. Fitch, F. J. & Miller, J. A. in *Earliest Man and Environments in the Lake Rudolf Basin* (eds Coppens, Y., Clark Howell, F., Isaac, G. Ll. & Leakey, R. E. F.) (University of Chicago Press, 1976).
39. Brock, A. & Isaac, G. Ll. in *Earliest Man and Environments in the Lake Rudolf Basin* (eds Coppens, Y., Clark Howell, F., Isaac, G. Ll. & Leakey, R. E. F.) (University of Chicago Press, 1976).
40. Hillhouse, J. W., Ndombi, J. W. M., Cox, A. & Brock, A. *Nature* **265**, 411–415 (1977).
41. Curtis, G. H., Drake, S., Cerling, T. & Hampel, J. *Nature* **258**, 395–398 (1975).
42. Hay, R. L. *Nature* **284**, 401 (1980).
43. Fuchs, V. E. *Phil. Trans. R. Soc.* **B229**, 219–274 (1939).
44. Watkins, R. & Williamson, P. G. *Proc. 8th Pan-African Congress of Prehistory and Quaternary Studies* (1979).

P. G. Williamson is at the Museum of Comparative Zoology, Harvard University.

This article was first published in *Nature* Vol. 293, pp. 437–443; 1981.

An uncensored page of fossil history

J. S. JONES

'A HISTORY of the world, imperfectly kept, and written in a changing dialect. Of this history we possess the last volume alone . . . Of this volume, only here and there a short chapter has been preserved; and of each page only here and there a few lines.' Thus Darwin on the imperfections of the fossil record, and thus his explanation—widely accepted until recently—of the absence from the record of the transitional forms expected on his view of the origin of species as a result of natural selection which acts 'solely by accumulating slight, successive, favourable variations, and can produce no great or sudden modification'.

In the preceding article, Williamson reports a page of the fossil history of the world that has been preserved more or less complete. He has studied a sequence of fossilized freshwater molluscs from the Turkana Basin in East Africa. In a series of fossil beds 400 m thick, which have yielded important human remains, there are preserved millions of shells of at least 19 species of snail from several genera. These deposits have remained more or less undisturbed since their formation, and the order of formation of each bed can be identified by referring to their relationship with accurately dated local geological features. This evolutionary series presents, as Williamson says, an unprecedented opportunity to study patterns of evolutionary change in a complete fossil record.

On first sight, these patterns are quite different from those expected on the theory of gradual evolution insisted upon by Darwin. In the 13 lineages common enough for detailed analysis, long periods of morphological stability are interrupted by fossil beds in which relatively rapid changes in shell shape take place. These newly evolved populations then persist unchanged through thick deposits before, in most cases, becoming extinct and being replaced by fossils resembling the ancestral forms (some of which persist to the present day). The intermediate forms between the ancestral and derived species occupy only a very small proportion of the evolutionary history of each lineage. The periods of transition coincide with each other in the various genera; including—remarkably enough—*Melanoides*, a taxon which is in its modern guise an obligate parthenogen. This apparent ability of an asexual species to evolve as rapidly as its sexual relatives casts some doubt on the many theories which claim that sexual reproduction increases evolutionary flexibility[1].

There is no doubt that these patterns of stability interrupted by change are real. In this they differ from recent claims that the fossil history of man shows similar patterns; these rest only on the incompleteness of the human fossil record[2].

Williamson suggests that the evolution of his snails conforms to a model of change in the fossil record which can be traced to the work of Cuvier in the eighteenth century; was later promoted by Goldschmidt[3], and has recently been resurrected as the 'punctuated equilibrium' model of Eldredge and Gould[4]. Evolution is seen as an essentially discontinuous process in which long periods of genetic stability are interrupted by 'genetic revolutions' which produce very rapid change, rather than as a system of gradual change leading to the origin of species by natural selection acting on minor differences in fitness among individuals. Darwin's often quoted claim that '*natura non fecit saltum*' is not correct; the absence of intermediate forms in the fossil record is a genuine result of the mechanisms of evolution, rather than an artefact of the accidents of preservation. To quote Williamson, the patterns of change in his fossil snail populations must indicate that 'speciation is a *qualitatively* different phenomenon from gradual intraspecific evolutionary change'. The darwinian theory of the origin of species, if not wrong, is at least incomplete.

Does this new information on a uniquely complete palaeontological series indeed mean that geneticists must revise their views of how species originate? Here we see an important difference in the way in which those dealing with fossils and those who experiment on living organisms assess the rate of evolutionary change; because of the differences in the time scales habitually encountered by the two groups, what is an instant of evolutionary time to a palaeontologist may appear almost an infinity to a geneticist.

Williamson shows that the intermediate forms in his fossil series existed for between 5,000 and 50,000 years, periods much shorter than those of evolutionary stability in each lineage. The living relatives of these snails have a generation interval of between six and twelve months[5-7], so that the morphological changes between the stable shell forms took on the average perhaps 20,000 generations to complete. To most geneticists, this interval seems more than sufficient to enable gradual changes to lead to morphological evolution as great as that described by Williamson. It is the equivalent of perhaps a thousand years in a *Drosophila* population cage, six thousand years in a mouse selection experiment, or 40,000 years when dealing with domestic animals such as dogs. Darwinian selection of the most conventional kind has often been used to accomplish dramatic genetic changes in morphology much more rapidly than this. In less than fifty generations of artificial selection on a base population it has been possible to produce selected lines of *Drosophila melanogaster* which have abdominal bristle numbers varying from three to eighty-five[8]; of mice whose body weights are as different as 13 g and 32 g[9], and of corn in which one selected line has an oil content of 15 per cent compared with only one per cent in that selected in the opposite direction[10]. In each case these changes have been achieved by gradualistic selection, and it is not necessary to postulate that new evolutionary mechanisms must be involved.

Reproductive isolation, the central element of speciation, can also be achieved by simple darwinian selection in a period much shorter than that assessed by Williamson and others as evidence for the inadequacy of darwinian theory. In *Drosophila melanogaster*, for example, gradualistic selection for preferential mating among mutants in a population cage can lead to considerable reproductive isolation within only 18 generations[11], and it is possible greatly to alter the degree of sexual isolation of *D. pseudoobscura* from its close relative *D. persimilis* by artificial selection in less than 20 generations[12]. In the same way, selection of different parts of an originally freely interbreeding *Drosophila* or house fly population for bristle number, climbing ability or the ability to tolerate insecticides can lead to the incidental evolution of considerable reproductive isolation within a few tens of generations[13-15]. Even *Drosophila* populations kept at different temperatures for five years in the laboratory evolve considerable reproductive isolation among themselves[16]. These are periods trivial in relation to those interpreted by Williamson and others as 'punctuations' in the fossil record which can only be explained by new evolutionary mechanisms.

The efficacy of gradual evolutionary change in producing genetic subdivision and the reduction of gene flow in natural populations subject to the forces of gradual selection is particularly well seen in plants. Some annual grasses have evolved an ability to grow on mines polluted by concentrations of copper high enough to kill populations not exposed to this selective agent. The grasses on the mine flower about a week earlier than do the surrounding non-tolerant populations and have an increased ability for self-fertilization. These genetic differences—which have evolved in the one hundred years since the mines were opened—are enough to lead to considerable reproductive isolation between mine populations and their ancestors in the nearby pastures[17].

Once evolutionary change in the fossil record—even in a record as well characterized as that unearthed by Williamson—is placed in the context of the known ability of living organisms to respond to the forces of classical darwinian natural selection, it becomes clear that it is not necessary to invoke evolutionary forces 'qualitatively different' from those emphasized by Darwin. Depending on the time scale to which the investigator is accustomed, one man's punctuated equilibrium may be another's evolutionary gradualism. Williamson describes an extraordinarily complete page in the history of evolution but its contents do not force us to change our views on the genetic mechanisms of the origin of species.

1. Maynard Smith, J. *The Evolution of Sex* (Cambridge University Press, 1978).
2. Cronin, J. E., Boaz, N. T., Stringer, C. G. & Rak, Y. *Nature* **292**, 113 (1981).
3. Goldschmidt, R. B. *The Material Basis of Evolution* (Yale University Press, 1940).
4. Eldredge, N. & Gould, S. J. *Models in Paleo-*

biology (ed. Schopf, T. J. M.) (Freeman, San Francisco, 1972).
5. Berry, A. J. & Kadri, A. B. H. *J. Zool., Lond.* **172**, 369 (1974).
6. Mackie, G. I. & Huggins, D. G. *J. Fish. Res. Bd. Can.* **33**, 1652 (1976).
7. Russell-Hunter, W. D. In *The Pulmonates* (ed.

Fretter, V. & Peake, J. F.) (Academic, New York, 1978).

8. Clayton, G. A., Morris, J. A. & Robertson, A. J. *Genet.* 55, 131 (1957).
9. Falconer, D. S. *Genet. Res. Camb.* 22, 29 (1973).
10. Woodworth, C. M., Leng, E. R. & Jugenheimer, R. W. *Agron. J.* 44, 60 (1952).
11. Knight, G. R., Robertson, A. & Waddington, C. H. *Evolution* 10, 14 (1956).
12. Koopman, R. F. *Evolution* 4, 135 (1950).

13. Thoday, J. M. *Proc. Roy. Soc.* B182, 109 (1972).
14. Hurd, I. E. & Eisenberg, R. M. *Am. Nat.* 109, 353 (1975).
15. King, J. C. *Cold Spring Harb. Symp. quant. Biol.* 20, 311 (1955).
16. Kilias, G., Alahiotis, S. N. & Pelecanos, M. *Evolution* 34, 730 (1980).
17. McNeilly, T. & Antonovics, L. *Heredity* 23, 205 (1968).

J. S. Jones is in the Department of Genetics and Biometry, University College London.

This article first appeared in *Nature* Vol. 293, pp. 427–428; 1981.

Morphological stasis and developmental constraint: real problems for neo-Darwinism

P. G. WILLIAMSON

THE preceding article[1] comments on a paper of mine[2] (reproduced on pp. 154–166) which summarizes a morphometric analysis of mollusc lineages from a Cenozoic sequence in Kenya. I made three key points: (1) all lineages exhibit morphological stasis for very long periods of time (3–5 Myr), (2) evolutionary change in each lineage is concentrated in relatively rapid speciation events (occurring over 5,000 to 50,000 years) and (3) the speciation events are accompanied by pronounced developmental instability in the transitional populations.

The first two points indicate that these lineages conform to the 'punctuated equilibrium' model for evolutionary change and the third is significant because the unusually complete Turkana Basin sequence offers the first fine-scale palaeontological documentation of the speciation process.

The preceding article considers only the second of these three points, making the uncontested observation that geneticists have succeeded in producing significant phenotypic changes in many populations over periods considerably shorter than those required for speciation events documented in the Turkana Basin mollusc sequence. The question of *rapidity* of speciation events is addressed but no attention is paid to the problems either of long-term morphological stasis or of developmental instability during speciation. Such comments are the standard but largely tangential criticisms advanced by many evolutionists against the punctuational model. The possibility of rapid change is freely admitted, but the significance of stasis, and the implications of developmental constraint for evolutionary process, are ignored. Since the critique (and others recently published)[3–5] seem to miss the most important issues raised by punctuated equilibrium theory in general, and my own work in particular, some comments seem in order.

As Gould and Eldredge[6] and many others have repeatedly pointed out, punctuated equilibrium is a theory about the *deployment of speciation in time*. It holds that many, perhaps most, metazoan fossil sequences show a characteristic pattern of morphological change through time: new species enter the record abruptly (in geological time), and persist with little significant change until extinction. Significant evolutionary change is, therefore, concentrated at speciation events.

For reasons that elude many of us, punctuated equilibrium, a theory of evolutionary tempo, has been conflated with Goldschmidtian macro-mutation, a theory of evolutionary mode. Thus, in the critique of my paper, I am lumped with Goldschmidt (and Cuvier of all people) as disciples of some peculiar (and non-existent) non-Darwinian evolutionary school. But punctuated equilibrium is compatible with much current neo-Darwinian thought. Eldredge and Gould[7], in their original formulation, relied upon the most orthodox version of Mayr's theory of allopatric speciation via peripheral isolates to account for the 'punctuational' pattern of morphological change in the fossil record. They considered the abrupt appearance of new species and their subsequent stasis, as well as the lack of intermediates between such stable lineages, to be compatible with (indeed, to flow from) Mayr's model. Mechanisms for rapid speciation that have been proposed subsequently (for example, the various models for 'chromosomal' speciation) are compatible with but not required by punctuated equilibrium, a theory of evolutionary tempo that is agnostic about modes of speciation, so long as they yield (as most standard models of speciation do) the punctuational pattern when translated into geological time.

It is not news to punctuationists that population geneticists can produce rapid phenotypic shifts in artificial selection regimes within a few generations—punctuationists and conventional neo-Darwinians are in complete agreement on this point. Why, then, are many punctuationists increasingly unhappy with conventional neo-Darwinian accounts of fundamental evolutionary process?

The principal problem is morphological stasis. A theory is only as good as its predictions, and conventional neo-Darwinism, which claims to be a comprehensive explanation of evolutionary process, has failed to predict the widespread long-term morphological stasis now recognized as one of the most striking aspects of the fossil record. The long-term morphological stasis noted by punctuationists in the fossil record is clearly mirrored by the relative morphological uniformity of most widely distributed modern species. As Ernst Mayr, the foremost student of geographical variation has written[8]: ' . . . it would . . . seem important to stress the basic uniformity of most continuously distributed species . . . The fact that (every taxonomist) . . . can identify individuals of a species . . . regardless of where in the range of the species they come from is further illustration of this phenomenon'. In a belated attempt to address the problem of morphological stasis, neo-Darwinists have invoked 'stabilising selection' (for example, Stebbins and Ayala[5]). But the wide range of environments presently exploited by extensively distributed but morphologically uniform modern species, and the long-term morphological stasis (up to 17 Myr) exhibited by many fossil lineages in fluctuating environments, strongly argues against the idea that simple stabilising selection is an adequate explanation for the phenomenon of morphological stasis. Accordingly, Mayr[8] explicitly invokes some form of developmental homeostasis, rather than stabilising selection, to explain the range-wide morphological stability of most modern species.

In the original formulation of punctuated equilibrium theory, it was suggested that temporal stasis, like the geographic stability noted by Mayr, was largely the result of some form of developmental constraint or homeostasis. In the absence of a comprehensive genetics of development, the mechanism for such homeostasis is obscure. But if some form of developmental homeostasis *is* at the root of morphological stasis, speciation must, by definition, involve the dismantling of homeostatic mechanisms pre-existing in the parental stock. The principal argument in my paper is that when speciation events occur in the Turkana Basin mollusc sequence, they are invariably accompanied by major developmental instability (that is, just such a dismantling of developmental homeostasis). The idea that morphological stasis is primarily a result of developmental homeostasis, and that speciation must therefore involve the temporary dismantling of such a homeostatic system, differs from most conventional neo-Darwinian assumptions about the way in which species arise. Most neo-Darwinists would agree with Darwin's statement that new species arise 'solely by accumulating slight, successive, favourable variations'. They believe that the intrapopulation micro-evolutionary changes observed in the *Drosophila* cage can be simply extrapolated into the differences between *Drosophila* species and Dipteran families. In this view of speciation, as Gould[9] says, there is a 'seamless continuum' a 'smooth extrapolation . . . from base substitution to the origin of higher taxa'. There is no suspicion here that the fundamental developmental constraints implied by long-term geographical and temporal stasis of species must be dismantled when new species arise. There is no suspicion that this radical reorganization of fundamental homeostatic mechanisms during speciation must involve a more radical overhaul of the phenotype than the steady 'march of metric means' seen in a *Drosophila* cage experiment.

Interestingly enough, the idea that disruption of developmental homeostasis, or constraints, is central to the speciation process is hardly new to the neo-Darwinian literature: Carson[10] has postulated 'open' and 'closed' genetic systems, the former involved in the allele-shuffling of minor adaptive adjustments within species populations, and the latter involved in the more profound regulatory and developmental changes during speciation. Levin[11] has pointed out the significance of developmental disruption and instability during the speciation process. But such suggestions have been largely ignored: the implied decoupling of micro-evolutionary change and macro-evolutionary phenomenon threatens the reductionist core of conventional neo-Darwinism.

The Turkana Basin sequence records a pattern of long-term stasis punctuated by rapid speciation accompanied by pronounced developmental instability. Jones suggests that this pattern requires no change in conventional views of the genetic mechanisms of the origin of species— despite the fact that this pattern is neither predicted by neo-Darwinism nor explicable in terms of its major tenets. Punctuationists suggest that it is time for conventional neo-Darwinism to address the important issues of morphological stasis and developmental constraint.

1. Jones, J. S. *Nature* **293**, 427 (1981).
2. Williamson, P. G. *Nature* **293**, 437 (1981).
3. Maynard Smith, J. *Nature* **289**, 13–14 (1981).
4. Maynard Smith, J. *London Review of Books*, 18th June–1st July (1981).
5. Stebbins, G. L. & Ayala, F. J. *Science* **213**, 967 (1981).
6. Gould, S. J. & Eldredge, N. *Paleobiology* **3**, 115 (1977).
7. Eldredge, N. & Gould, S. J. in *Models in Paleobiology* (ed. Schopf, T. J. M.) (Freeman, Cooper, San Francisco, 1972).
8. Mayr, E. *Populations, Species, and Evolution* (Harvard University Press, Cambridge, 1963).
9. Gould, S. J. *Paleobiology* **6**, 119 (1980).
10. Carson, H. L. *Am. Nat.* **109**, 83 (1975).
11. Levin, D. A. *Am. Nat.* **104**, 343 (1970).

P. G. Williamson is at the Museum of Comparative Zoology, Harvard University.

This article first appeared in *Nature* Vol. **294**, pp. 214–215; 1981.

Intense natural selection in a population of Darwin's finches (Geospizinae) in the Galápagos

PETER T. BOAG AND PETER R. GRANT

Survival of Darwin's finches through a drought on Daphne Major Island was nonrandom. Large birds, especially males with large beaks, survived best because they were able to crack the large and hard seeds that predominated in the drought. Selection intensities, calculated by O'Donald's method, are the highest yet recorded for a vertebrate population.

THERE are few well-documented examples of natural selection causing avian populations to track a changing environment phenotypically. This is partly because birds meet environmental challenges with remarkable behavioral and physiological flexibility[1], partly because birds have low reproductive rates and long generation times, and partly because it has been difficult for ecologists to quantify corresponding phenotypic and environmental changes in most field studies. In this report we demonstrate directional natural selection in a population of Darwin's finches and identify its main cause.

We studied Darwin's medium ground finch (*Geospiza fortis*) on the 40-ha islet of Daphne Major, the Galápagos, from July 1975 to June 1978. Each of more than 1500 birds was color-banded and measured for seven external morphological characters[2]. Continuous records were kept of the banded birds and of rainfall. Each year during the breeding season (January to May) we banded nestlings and compiled nest histories. Three times a year (before, during, and after the breeding season) we collected the following data: (i) the number of seeds of each plant species in 50 randomly chosen 1.0-m² quadrats; (ii) a standardized visual census of finches over the entire island; and (iii) a minimum of 100 point records of feeding behavior, accumulated by noting food items eaten by banded birds encountered during non-systematic searches[2].

During the early 1970s Daphne Major received regular rainfall, resulting in large finch populations and food supplies[2]. From December through June in 1976 and 1978 we recorded rainfalls of 127 and 137 mm, respectively—sufficient for abundant production of plants, insects, and finches. However, in 1977 only 24 mm of rain fell on Daphne Major during the wet season[3,4]. *Geospiza fortis* did not breed at all in 1977 and suffered an 85 percent decline in population (Fig. 1A). The decline was correlated with a reduction in seed abundance ($r = .86$, $P < .01$) (Fig. 1B). Seeds form the staple diet of *G. fortis*, particularly in the dry season, when other plant matter and insects are scarce[2].

Between June 1976 and March 1978, the mortality, and possibly emigration[5], of *G. fortis* was nonrandom with respect to age, sex, and phenotype. Only one of 388 *G. fortis* nestlings banded in 1976 survived to 1978, and while the sex ratio was roughly equal in 1976, it had become skewed to six males to one female in 1978. Most significantly, the birds surviving into 1978 were considerably larger than those that disappeared (Fig. 1C). We use principal component 1[6] as an index of overall body size because here, as in other avian studies[7], it explains a substantial portion (67 percent) of the phenotypic variance in the *G. fortis* population and has consistently high, positive correlations with the morphological variables it summarizes. The change is most obvious in the plot including all birds because it incorporates the changing sex ratio (most of the morphological characters are 4 percent larger in males than in females) and perhaps a small age effect, although all birds less than 12 weeks old were excluded from the analysis.

Small seeds declined in abundance faster than large ones, resulting in a sharp increase in the average size and hardness of available seeds (Fig. 1D). There was a corresponding change in feeding behavior. In May 1976 only 17 percent of feeding was on medium or large seeds [size-hardness index $\sqrt{DH} \geq 1.0$][8], while in May 1977 49 percent of feeding was on such seeds. During the present and related studies[2], large birds ate larger seeds than smaller birds, suggesting that small birds disappeared because they could not find enough food. For example, in a quantitative test of size-related feeding behavior, 198 birds that were only recorded eating seeds with a size-hardness index < 1.0 were significantly smaller than another 121 birds that routinely ate seeds with size-hardness indices ranging from 1.0 to 8.7[8]. In 1977, during the normally lush wet season, larger birds fed heavily on seeds extracted from the large, hard mericarps of *Tribulus cistoides* ($\sqrt{DH} = 8.68$), a food item ignored by almost all birds in earlier years[2]. Many finches failed to molt that year, and their condition gradually deteriorated. Small birds fed heavily on *Chamaesyce* spp., the only producer of small seeds in 1977, and as result their plumage often became matted with the latex of this euphorb. Several dead birds were found with completely bald heads from feeding on *Chamaesyce* and from digging in the soil for seeds. Such plumage loss may have led to increased energy loss during the cool nights of the dry season. The dependence of the finches on a declining seed supply ceased at the end of 1977, when *Opuntia* cactus began flowering and all birds fed heavily on its pollen and nectar[2].

It is reasonable to infer natural selection from the greater survival of large birds because about 76 percent of the variation in the seven morphological measurements and in principal component 1 scores is

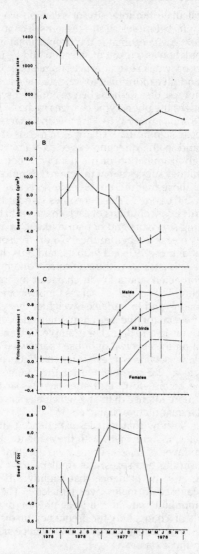

FIG. 1. *Temporal changes in finch numbers, seed abundance, morphology, and average seed size on Daphne Major. (A) Population estimates (means ± 95 percent confidence limits) derived from a Lincoln index based on regular visual censuses of a marked population. (B) Estimates of seed abundance [means ± standard errors (S.E.)], excluding two seed species never eaten by any Galápagos finches. (C) Principal component 1 scores (means ± S.E.) for birds alive in each sample period, with coefficients calculated from the combined sample of all birds measured. (D) Estimates of the average $\sqrt{DH}$ index (means ± S.E.) of edible seeds available in each study period[8].*

heritable[3,9]. To calculate the intensity of selection we use O'Donald's method[10], $\Delta\bar{w}/\bar{w} = (\bar{w}$ before selection—$\bar{w}$ after selection)$/\bar{w}$ before selection $= V_w/\bar{w}^2$, *where $\Delta\bar{w}/\bar{w}$ estimates the proportional increase in mean fitness of the population as a result of selection and V_w is the variance in fitness.* O'Donald provides several functions relating fitness to phenotypic characters and gives formulas for calculating $\Delta\bar{w}/\bar{w}$ from the four moments of phenotypic distributions before and after selection.

Table 1 summarizes the phenotypic changes in the *G. fortis* population between June 1976 and January 1978. Changes in variance were small and none was statistically significant[11]. Changes in means of most characters were significant and in the direction expected if larger birds survived best[12]. A thorough examination of the data with both univariate and multivariate techniques suggests that the main differences between birds that survived and those that did not were in body size and bill dimensions, particularly bill depth[8]. Table 1 includes standardized coefficients that show the relative contributions of each character to the discriminant functions separating survivors and nonsurvivors. Our analysis includes only adult finches measured before the 1976 dry season; the 1978 survivors are a subset of those 1976 individuals, and thus the 1978 range for any given variable falls entirely within the corresponding 1976 range.

Because selection acted primarily on character means, we assume a linear fitness function[10]. The highest values for $\Delta\bar{w}/\bar{w}$ are observed in the discriminant functions and in variables weighted heavily by the functions (Table 1). Several of the selection intensities are considerably greater than any published to date. For example, O'Donald[10] reanalyzed H. Bumpus's data on the survival of house sparrows (*Passer domesticus*) during a particularly severe winter storm, and concluded that such values as the $\Delta\bar{w}/\bar{w} = .255$ he obtained for the change in discriminant score between the before-storm and after-storm sparrow samples indicated selection 'more intense than any which has since been observed acting on particular quantitative characters'[10].

Table 1 and Fig. 1 show that females experienced stronger selective mortality than males, in agreement with the evidence that the sex ratio became skewed in favor of males. There is no question that the overall effect of selection in the two sexes was similar: larger individuals survived best. There is some evidence that slightly different aspects of 'largeness' were favored in males over females[13]. The results for the combined population illustrate how a large phenotypic shift can occur both as the result of changes in the frequency of discrete classes of individuals (males and females) and in the average measurements of individuals within those classes.

Our data provide a link between a specific environmental factor (size of available food) and phenotypic tracking of the environment. Others have consistently encountered difficulty in identifying the relation between complex and often rather small changes in body size and shape and general environmental parameters, such as temperature[14]. Because of the high correlations between the seven characters we examined, it is difficult to specify the precise target of selection; univariate selection intensities and discriminant coefficients presented in Table 1 and calcu-

TABLE 1. *Characteristics of finches surviving the 1977 drought. The sample measured before selection includes all mature G. fortis measured up to the end of May 1976. The sample measured after selection is the subset of the first sample still present on Daphne Major in March 1978. Standardized discriminant function coefficients (SDFC's) reflect the relative contribution of each univariate character to the discriminant function separating survivors and nonsurvivors. Principal component 1 is an index of body size and principal component 2 reflects bill pointedness: both are presented as standardized variables[7]. Separate discriminant functions[11] were used to distinguish between survivors and nonsurvivors in the combined sample of birds emphasized here and in smaller samples of males only and females only, for which the discriminant and principal component scores alone are given here. The mean discriminant scores are unstandardized and, because different functions were used, scores of the three groups are not comparable. Values for Δw̄/w̄ give the proportionate increase in mean fitness as a result of selection, assuming a linear fitness function[10].*

Variable	Sample size		Means		Variances		SDFC	Δw̄/w̄
	Before	After	Before	After	Before	After		
All birds								
Weight (g)	642	85	15.79	16.85	2.37	2.43	0.45	.49
Wing chord (mm)	642	85	67.71	69.22	5.89	5.01	0.35	.39
Tarsus length (mm)	642	85	18.76	19.11	0.57	0.49	0.13	.22
Bill length (mm)	641	85	10.68	11.07	0.55	0.57	0.14	.23
Bill depth (mm)	642	85	9.42	9.96	0.68	0.66	0.45	.44
Bill width (mm)	641	85	8.68	9.01	0.36	0.34	-0.56	.31
Bill length at depth of 4 mm	642	85	3.55	3.41	0.08	0.08	-0.35	.24
Principal component 1	640	85	0.00	0.73	1.06	0.95		.50
Principal component 2	640	85	0.01	-0.13	0.97	0.99		.02
Discriminant function	640	85	-11.55	-12.41	1.13	0.99		.66
Males only								
Principal component 1	198	56	0.54	0.98	0.92	0.76		.21
Discriminant function	198	56	7.35	6.87	0.75	0.64		.31
Females only								
Principal component 1	66	15	-0.21	0.32	0.97	0.96		.28
Discriminant function	66	15	-2.77	-3.71	1.57	1.81		.57

lated in the similar analyses of separate male and female groups[13] suggest that weight and bill dimensions are most important. In addition to the relation between bill morphology and changes in the food supply, it is likely that there were additional indirect selection pressures operating on, for example, body size for reasons associated with energetics[15] and dominance behavior[16]. Furthermore, it is likely that a different set of selection pressures operates when food is abundant and population size is increasing, thus giving rise to oscillating directional selection[2].

Our results are consistent with the growing opinion among evolutionary ecologists that the trajectory of even well-buffered vertebrate species is largely determined by occasional 'bottle-necks' of intense selection during a small portion of their history[17]. More specifically, given the many small, isolated, relatively sedentary, and morphologically variable populations of Darwin's finches[18, 19] and the high spatial[2, 19] and temporal[4] variability of the Galápagos, this type of event provides a mechanism for rapid morphological evolution. Occasional strong selection of heritable characters in a variable environment may be one of the keys to explaining the apparently rapid adaptive radiation of the Geospizinae in the Galápagos[18, 20].

1. Slobodkin, L. B. in *Population Biology and Evolution*, Lewontin, R. C. Ed. (Syracuse Univ. Press, Syracuse, N.Y., 1968), pp. 187–205; Selander, R. K. & D. W. Kaufman, *Proc. Natl. Acad. Sci. U.S.A.* **70**, 1875 (1973).

2. Grant, P. R., Smith, J. N. M., Grant, B. R., Abbott, I. J. & Abbott, L. K. *Oecologia (Berlin)* **19**, 239 (1975); Smith, J. N. M., Grant, P. R., Grant, B. R., Abbott, I. J. & Abbott, L. K. *Ecology* **59**, 1137 (1978); Grant, P. R. & Grant, B. R. *Oecologia (Berlin)* **45**, 55 (1980); *Ecol. Monogr.* **50**, 381 (1980).

3. Boag, P. T. thesis, McGill University, Montreal (1981); Grant, P. R., Grant, B. R., Smith, J. N. M. Abbott, I. J. & Abbott, L. K. *Proc. Natl. Acad. Sci. U.S.A.* **73**, 257 (1976).

4. Grant, P. R. & Boag, P. T. *Auk* **97**, 227 (1980).

5. Grant, P. R., Price, T. D. & Snell, H. *Not. Galápagos* **31**, 22 (1980). This article reports the only documented finch emigration from Daphne Major, involving two juvenile *G. scandens*. There is indirect evidence that most of the missing *G. fortis* died on Daphne Major; the measurements of 38 *G. fortis* banded before June 1976 and found dead on Daphne in 1977 and early 1978 are statistically indistinguishable from the other birds missing from the postselection population but not found. The individuals found dead also were significantly smaller than the 1978 survivors, with both the entire sample and males and females separately showing patterns similar to those described for the entire set of missing birds[12].

6. Principal components were extracted from the covariance matrix of the seven log-transformed variables, with all birds combined. The first component explained 67 percent of the total variance and had large correlations with the seven original variables; following the sequence used in Table 1, these component-character correlations were .88, .67, .60, .85, .94, .93, and −.49. Component 2 explained a further 16 percent of the variance and was strongly correlated with bill length at a depth of 4 mm ($r = .87$), followed by bill length ($r = .30$), with other characters showing low correlations of mixed signs. Correlations with length at a depth of 4 mm are reversed because the character is necessarily smaller in larger birds. Other analyses with these finches confirm that principal components 1 and 2 usually reflect overall body size and bill 'pointedness,' respectively, whether based on covariance or correlation matrices.

7. Blackith, R. E. & Reyment, R. A. *Multivariate Morphometrics* (Academic Press, New York, 1971); Gibson, A. R., Gates, M. A. & Zach, R. *Can. J. Zool.* **54**, 1679 (1976); Ricklefs, R. E. & Travis, J. *Auk* **97**, 321 (1980); Wiens, J. A. & Rotenberry, J. T. *Ecol. Monogr.* **50**, 287 (1980).

8. The size-hardness index is the geometric mean of the depth (D) of a seed species in millimeters and its hardness (H) in newtons. In Fig. 1D, average $\sqrt{DH}$ values are obtained by weighting the mean number of seeds (of each species) per square meter by the average seed mass and $\sqrt{DH}$ for that species[2]. The tests of hard-versus soft-seed feeders consisted of univariate t-tests on all seven morphological variables; in all cases birds feeding on large seeds were significantly larger ($P < 0.001$) than those feeding only on small seeds. The seven-variable multivariate analysis of variance between the two groups was also highly significant

$[F(7,311) = 7.94, P < .0001]$. The standardized coefficients of the discriminant function separating the two feeding groups weighted bill depth most heavily (.80), followed by wing chord (.57), with all other variables having coefficients under .25. This underlines the link between bill depth and feeding behavior, which persisted among the survivors at the end of 1977 (Grant, P. R. *Anim. Behav.* in press).

9. Boag, P. T. & Grant, P. R. *Nature (London)* **274**, 793 (1978).
10. O'Donald, P. *Theor. Popul. Biol.* **1**, 219 (1970); *Evolution* **27**, 398 (1973).
11. Separate linear discriminant functions were computed for males alone, for females alone, and for both combined with mature birds of uncertain sex, each maximizing the distance between the centroids of survivors and nonsurvivors in that group. Unstandardized discriminant scores were calculated for each bird by using the appropriate equation. The variances for the seven original variables, the first and second principal components, and the discriminant function of birds that survived from June 1976 to March 1978 were compared with those of birds that disappeared. The ten comparisons were made for males, for females, and for the combined group; in 21 of the 30 comparisons the selected group was less variable, but none of the 30 F-tests approached significance.
12. We computed t-tests for the 30 comparisons detailed in (11), again contrasting survivors and nonsurvivors to maintain sample independence. After the 1977 drought, males were significantly larger ($P < .01$) in all variables except wing chord, tarsus length, and principal component 2. Females were significantly larger ($P < .05$) in all variables except weight and tarsus length, with prinipal component 2 on the borderline ($P = .066$). The combined group was significantly larger ($P < .001$) in all variables except principal component 2.
13. The three largest standardized coefficients of the discriminant function for males alone were for bill depth (1.00), weight (.85), and bill width (.56), and the three largest $\Delta \overline{w}/\overline{w}$ values for male univariate characters were again for bill depth (.22), weight (.20), and bill width (.15). The corresponding results for the female group were different; the largest standardized discriminant function coefficients were for bill length (-1.21), bill length at a depth of 4 mm (.90), and bill depth (.55), and the largest $\Delta \overline{w}/\overline{w}$ values were for bill length at a depth of 4 mm (.40), bill depth (.25), wing chord (.22), bill length (.20), and bill width (.20).
14. James, F. C. *Ecology* **51**, 365 (1970); Grant, P. R. *Syst. Zool.* **21**, 23 (1972); Johnston, R. F., Niles, D. M. & Rohwer, S. A. *Evolution* **26**, 20 (1972); Lowther, P. E. *ibid.* **31**, 649 (1977); Baker, M. C. & Fox, S. F. *Am. Nat.* **112**, 675 (1978); Johnson, D. M. *et al.*, *Auk* **97**, 299 (1980).
15. Kendeigh, S. C. *Am. Nat.* **106**, 79 (1972).
16. Fretwell, S. D. *Bird-Banding* **40**, 1 (1969); Morse, D. H. *Am. Nat.* **108**, 818 (1974); Baker, M. C. & S. F. Fox, *Evolution* **32**, 697 (1978).
17. J. A. Wiens, *Am. Sci.* **65**, 590 (1977).
18. Lack, D. *Darwin's Finches* (Cambridge Univ. Press, Cambridge, 1947).
19. Bowman, R. I. *Univ. Calif. Berkeley Publ. Zool.* **58**, 1 (1961); Lowther, P. E. *ibid.* **31**, 649 (1977). Abbott, I. J., Abbott, L. K. & Grant, P. R. *Ecol. Monogr.* **47**, 151 (1977).
20. Bailey, K. *Science* **192**, 465 (1976).
21. This work was supported by the National Research Council of Canada and the Frank M. Chapman Fund of the American Museum of Natural History and was carried out with the permission of the Dirección General de Desarrollo Forestal, Quito, Ecuador. The Charles Darwin Research Station arranged logistics in the Galápagos. We thank B. R. Grant, E. Green, D. Nakashima, L. M. Ratcliffe, and R. Tompkins for assistance in the field and G. Bell, B. R. Grant, T. D. Price, L. M. Ratcliffe, and the reviewers for their advice on the manuscript. This is contribution 309 from the Charles Darwin Foundation.

Peter T. Boag and Peter R. Grant prepared this article at the Department of Biology, McGill University, Montreal. Peter T. Boag is in the Department of Biology, Trent University, Peterborough, Ontario and Peter R. Grant in the Division of Biological Sciences, University of Michigan.

CHAPTER SIX

THE EVOLUTION OF BEHAVIOUR

DARWIN's books, *The Expression of the Emotions in Man and Animals*, and *The Descent of Man and Selection in relation to Sex* demonstrate his interest in the evolution of behaviour, and also suggest that his interest was inseparable from his curiosity about the nature and evolution of man. Current work on the subject goes back to the ethologists Lorenz and Tinbergen, but in recent years the ethological approach has been supplemented by a demand for a more rigorous selective explanation for the evolution of particular traits, which originates with Hamilton's 1964 papers on the evolution of social behaviour.

The paper by Sherman is essentially a test of Hamilton's ideas. For a ground squirrel to give an alarm is potentially dangerous. How, then, can a Darwinist account for the giving of alarms? The proposal is that a gene which causes its carrier to perform an act which puts it at risk may nevertheless increase in frequency if the result of that act is to help relatives of the actor, who may carry identical copies of the gene. To test this process of 'kin selection' it becomes necessary to know the genetic relationship between interacting individuals.

An alternative reason why animals may help each other is as follows. Two animals may cooperate because each is better off cooperating than it would be if it defected. Such interactions are termed 'mutualistic'. Most analyses of social behaviour suggest that both kin selection and mutualism are relevant. Particular difficulties arise if cooperative acts are not synchronous; that is, if A helps B today and B helps A tomorrow. Trivers argued that such 'reciprocal altruism' could evolve provided that there are repeated opportunities for interaction, and that individual recognition and learning are possible. Thus the stability of reciprocal altruism against 'cheating' depends on the ability of an animal whose help is not reciprocoated to stop helping. Packer's paper on Olive Baboons shows that, at least in highly sophisticated primates, Trivers' mechanism can work.

In his reciprocal altruism paper, Trivers noted that the problem can be treated by game theory, crediting the suggestion to Hamilton. The paper by Axelrod and Hamilton printed here pursues this possibility further. They consider the Prisoner's Dilemma game, played repeatedly between a pair of opponents. Game theorists have known for some time that the strategy of 'Tit for Tat'—start by cooperating, and then do in each game what your opponent did in the last—is a solution to the repeated Prisoner's Dilemma game. Axelrod and Hamilton consider the game in an evolutionary context, and show that Tit for Tat is uninvadeable—that is, it is an 'Evolutionarily Stable Strategy', or ESS. They go on to discuss the possible biological significance of this finding.

The idea of sexual selection goes back to Darwin, but remarkably little work has been done on the topic until quite recently. It often turns out that mating is not 'random'; instead, some kinds of individuals mate more often, or are more likely to mate with particular kinds of partner. Darwin envisaged two possible selective mechanisms—female choice and male-male competition. These are not always easy to distinguish, even in theory. If males fight, and females passively accept the winner, then there is no element of female choice. We can reasonably speak of choice if females have evolved some method of discrimination causing them to mate with some kinds of conspecific males in preference to others (rather than merely causing them to mate with conspecifics in preference to males of other species). Such discrimination will not evolve unless females that discriminate have more offspring, or fitter offspring, than females that mate randomly. The paper by Partridge demonstrates non-random mating, and also that females given an opportunity of exercising choice have fitter offspring. It is true that the fitness differences she finds are small, but they are quite large enough to be decisive in evolutionary time. However, it is not clear whether mating is non-random because females have evolved an ability to discriminate, or because some males court more vigorously or frequently.

The papers by Sherman, Packer and Partridge have in common that they confront a theory of how selection acts with experimental field observation on a particular species. An alternative approach, illustrated in the paper by Harcourt et al., is to test some selective hypothesis by using comparative data on many species. Thus it turns out that not only do larger primates have larger testes (it would be strange if they did not), but, for animals of a particular size, the testis is larger in those species in which a female is likely to mate with more than one male during a single oestrus, than it is in monogamous species or in species in which a single male holds a 'harem' of several females. The reason, presumably, is that in the former species sperm competition determines which male becomes the father.

One reason for being interested in these particular data is that we are primates, and that our testis size lies (slightly) below the line of best fit, suggesting that our ancestors were monogamous or harem-holding rather than promiscuous. Martin and May, in their article, go further, to point out that in monogamous primates there is never a substantial difference in size between the sexes. Since such a dimorphism exists in our own species, they suggest that the most likely breeding system for our ancestors was that of a single male bonded to several females. The conclusion should be treated with caution. Comparative data can show, for example, that sperm competition is one of the causes of large testis size, and that polygyny is one of the causes of sexual dimorphism in size. However, such data cannot show that these are the only causes, and one must therefore be careful in drawing conclusions about particular species, particularly an ecologically aberrant species like man. Nevertheless, these are the best indications that comparative biology can provide at the moment.

Nepotism and the evolution of alarm calls

PAUL W. SHERMAN

ALARM calls, vocalizations that alert other animals to impending danger, give the appearance of altruism. Identifying the function of the alarm calls of any species has proved difficult, both because predation is rarely seen in the field[1] and because individual identity of and kinship among members of prey species are usually unknown. Moreover, members of many species give several different, predator-specific alarm calls.

During a 3-year field study, I investigated the function of the alarm call that Belding's ground squirrels (*Spermophilus beldingi*, Rodentia: Sciuridae) give when a terrestrial predator approaches. Because the ground squirrel population that I studied contains individually marked animals of known age, among which familial relationships through common female ancestors are also known, discriminating among several hypothesized advantages of giving alarm calls is for the first time possible. A disadvantage of calling is also demonstrated. My investigation indicates that assisting relatives, nepotism, is the most likely function of the ground squirrels' alarm call; this result implicates kin selection[2] in the evolution of a behavior that, because it may involve risks to the alarm caller's phenotype, appears to be altruistic.

FUNCTIONS OF ALARM CALLS

Individuals may benefit from giving alarm calls in any of several contexts, because alarm calls may result in one or more of the following six effects.

(1) *Diversion of predators' attention to other prey.* This hypothesis would be implicated if, in the absence of cover, alarm calls or screams from captured individuals stimulate aggregation[3], group mobbing[4,5], or pandemonium[5-7]; or, if the prey are already hidden, alarm calls cause them to behave in a manner that would enhance their crypticity[6,7]. Observations suggesting that 'ventriloquial' alarm calls occur that increase the jeopardy of others[8] or that callers mislead or manipulate conspecifics so as to increase their own safety[6] would also support this hypothesis for the species and call at issue.

(2) *Discouragement of predator pursuit.* By calling, potential prey may

reduce the likelihood and costs of attacks on themselves, if calls cause predators to terminate pursuits. For example, fleet and elusive prey might discourage predators by indicating to the predators that they have been seen and that the advantage of surprise has thus been removed[9]. Sudden or erratic changes in prey behavior as well as alarm calls may startle or momentarily confuse predators, and may indicate to them that an attack is unlikely to succeed[10]. In addition, poisonous prey might signal their distastefulness by giving an alarm call[4]. Under this hypothesis, callers gain by indicating to a predator that it has been detected or that the probability of a successful or profitable attack is low. This second hypothesis would thus be implicated if predators consistently turn away from or suddenly release callers, regardless of the presence, proximity, or behavior of other suitable prey.

(3) *Alerting relatives.* Callers may gain by having placed themselves in some jeopardy if kin are thereby consistently warned[2,11,12]. Captured individuals might also give distress (alarm) calls in this context, thereby soliciting assistance from relatives[4] or else warning them to flee or to hide. Under this, the third hypothesis, year-round alarm calls must be associated with the continuous presence of relatives (compare Williams[12] p. 206). If alarm calls are given during only part of the year, they must coincide with proximity of kin. For a given species, this hypothesis would be strongly supported if individuals with relatives living within earshot call more frequently than do conspecifics without them.

(4) *Helping the group.* Alarm calling might spread by a process of between-group selection, either if (i) prey populations are composed of small, genetically isolated demes[13] or if (ii) between periods of dispersal and panmixia, prey populations are sedentary and composed of isolated aggregations of individuals that are similar to each other in their propensity to call[14,15]. Then either (i) the persistence of groups must be proportional to the percentage of callers within them and groups containing more callers must recolonize areas left vacant by the extinction of groups containing fewer callers[13] or else (ii) temporary aggregations of sedentary individuals must produce dispersing young in proportion to the percentage of callers within each aggregation[14]. In both cases (i) and (ii), unlike the case where the nepotism hypothesis (that is, the third hypothesis) is applicable, fully or partially isolated groups of conspecifics must be identifiable[13-16], and these groups must differ in the proportion of alarm callers versus noncallers. If identifiable groups exist and if between-group differences in percentage of callers are demonstrable, the familial relationships among group members must then be considered[16,17] because between-group differences in the percentage of callers could be brought about by the association of either family members or of nondescendants. If the former, the differential reproduction of such groups is most appropriately analyzed in terms of kin selection[16,17] (but see 18). If the latter, hypothesis 4 can be distinguished from hypothesis 3.

(5) *Reduction of the likelihood of later attacks by the same predator.* If predators become better at hunting similar prey with experience or if they return to hunt near sites of previous successful kills [for examples, see 19], alarm

callers may benefit by warning conspecifics if by so doing they deny predators sustenance and a search image[20]. Hypothesis 5 implies that the phenotypic risk of calling is at least lower than the danger of being surprised during a later hunt by the same predator. The hypothesis requires that predators are more often successful in populations without alarm callers than in populations containing them. Hypothesis 5 does not require a particular population structure or familial relationship among callers and those warned. If predators return to sites of previous successful kills, hypothesis 5 predicts that the most sedentary individuals should call most frequently, because they will be in jeopardy from returning predators more often than less sedentary conspecifics.

(6) *Warning of others likely to reciprocate*. If individual callers and listeners associate long and consistently enough for them to exchange risks associated with alerting each other and benefits accompanying being alerted, alarm calling may spread on the basis of reciprocity[20]. As proposed by Trivers, this hypothesis assumes that callers and warned individuals are either distantly related or unrelated[20]; however, reciprocity may also occur among related conspecifics (21; see also 15), complicating efforts to contrast hypotheses 6 and 3. Hypothesis 6 would be supported if the likelihood of calling increases directly with the probability of warning reciprocators or if this likelihood decreases with the probability of warning nonreciprocators [for a possible example of reciprocity among primates, see 22].

Under hypotheses 1 and 2, alarm calling is favoured because of benefits to the caller's phenotype. Under hypotheses 3 to 6, alarm calls are phenotypically but not genotypically altruistic (21, p. 336).

STUDY AREA AND STUDY ANIMAL

During the summers of 1974 through 1976, ten different field assistants (three in 1974, five in 1975, and five in 1976) and I studied the responses of Belding's ground squirrels (Fig. 1) to terrestrial predators at Tioga Pass Meadow, in the Sierra Nevada mountains of California[23]. Ground squirrels in the study population have been permanently marked yearly since 1969: between 1969 and 1973, M. L. Morton and his students individually toe-clipped 731 of them; from 1974 to 1976 my assistants and I double-ear-tagged another 1135, including the 451 young from 101 complete litters. Therefore, exact ages (up to 8 years) of and familial relationships through common female ancestors among groups of ground squirrels are known. Most animals were marked with human hair dye for visual identification at a distance, and their burrows were marked with stakes and painted rocks.

During 3082 hours of observation, members of five species of terrestrial predators and marked ground squirrels of known age were seen simultaneously 102 times: long-tailed weasels (*Mustela frenata*) 67 times, badgers (*Taxidea taxus*) 11 times, dogs (*Canis familiaris*) unaccompanied by humans 11 times, coyotes (*Canis latrans*) 10 times, and pine martens (*Martes americana*) 3 times. On these occasions nine ground squirrels (six

FIG. 1. *Belding's ground squirrel at Tioga Pass, Mono County, California.*

adults and three juveniles) were killed (that is, one was killed every 342 observation hours): two by pine martens, three by coyotes, and four by long-tailed weasels. I use these observations to discriminate among hypotheses 1 to 6 for this species' alarm call.

Belding's ground squirrels are diurnal rodents that inhabit alpine and subalpine meadows in the Far West[24, 25]. At the study area, elevation 3040 meters, they are active from May through September, and they hibernate the rest of the year[23]. Although conspecific ground squirrels interact daily, they do not group their burrows into circumscribed aggregations nor do they produce young synchronously as do colonial species such as black-tailed prairie dogs (*Cynomys ludovicianus*)[26-28].

Like many other terrestrial sciurids[29,30], Belding's ground squirrels give a segmented alarm call in the 4- to 6-kilohertz range when a predatory mammal approaches them (Fig. 2); by contrast they give a single-note, high-pitched whistle to aerial predators (31; see also 32). Their alarm call to terrestrial predators is easily localized by humans, perhaps because of certain acoustical properties of the sound[33] (Fig. 2) and because individuals usually call repeatedly [$\bar{X} \pm$ standard error (S.E.) = 27.8 ± 3.8 calls per individual per predator appearance, with $N = 13$; $\bar{X} = 6.1 \pm 1.3$ minutes of calling per individual per predator appearance, $N = 16$], even after a predator has apparently disappeared ($\bar{X} = 3.7 \pm 0.9$ minutes of calling per individual, after the predator disappeared from an observer's view; $N = 19$). Vigorous vibrations of chest cavities of calling ground squirrels and their open mouths enhanced our ability to determine callers' identities, even when several animals were close together. Eighty-two times ground squirrels gave calls that sounded like alarm calls (that is, Fig. 2) when no predator was seen. Because these calls might not have been predator-related, I report here only behavior taking place on the 102 occasions when predators and ground squirrels were simultaneously seen, regardless of whether or not alarm calls were heard. For Tables 2 and 3 and Fig. 3, I combined data from appearances of all five species of predatory mammals after determining that neither the proportions of sex and age categories of ground squirrels present when a predator appeared (Fig. 3) nor the percentage of animals that called differed among predator species (all $P \geq .1$, two-tailed G statistics).

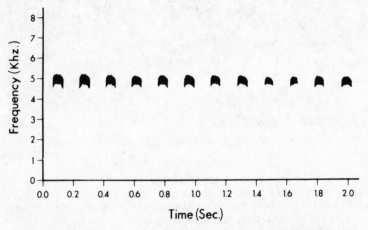

FIG. 2. *Sound spectrogram of the alarm call that Belding's ground squirrels give when predatory mammals appear. No frequency harmonics between 6 and 16 khz were found. Frequency is given in kilohertz and time in seconds.*

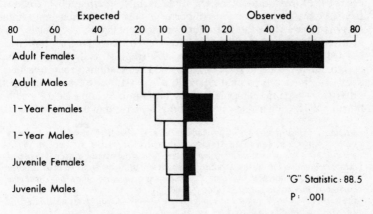

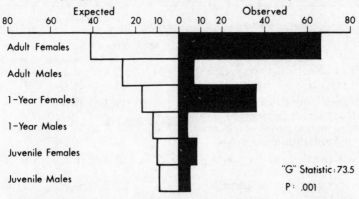

FIG. 3. *Expected and observed frequencies of alarm calling by various sex and age classes of Belding's ground squirrels. 'Expected' values were computed by assuming that animals call randomly, in direct proportion to the number of times they are present when a predatory mammal appears. The overall significance of both comparisons is largely due to females calling more often than 'expected' and males calling infrequently. Data are from 102 interactions between ground squirrels and predators (1974–76).*

POPULATION STRUCTURE AND MATING SYSTEM

At Tioga Pass Meadow, the average genetic relatedness among female ground squirrels inhabiting any small area is high as a result of common ancestry. As in several other terrestrial sciurids[34,35], females successfully rearing young are sedentary between years, and daughters mature and breed near their birthplaces until they die or disappear from the study area. In contrast to their sisters (Table 1), males permanently emigrate from the area where they were born, usually before their first winter hibernation[36]. Males do not return to their natal area to copulate, and brothers do not aggregate elsewhere. Seven males born in 1974 were sexually active for the first time in 1976 (that is, as 2-year-olds), and mated 422.0 ± 89.8 m ($X \pm$ S.E., $N = 11$ copulations) from their natal burrows; the brothers' matings took place 341.3 ± 107.6 m from each other ($N = 6$ pairs of copulations by brothers). By contrast, 12 females born in 1974, each a sister of one of the 2-year-old males, mated $43.2 \pm$

TABLE 1. *Within-family sexual asymmetries in emigration distances among Belding's ground squirrels at Tioga Pass Meadow, California. For females, the home burrow is either the one from which their offspring emerged or, if their young died or disappeared before emergence, the burrow to which they carried nesting material and in which they spent the nights at about the time the young were emerging. For males, the home burrow is the one to which they carried nesting material and in which they spent the nights at about the time the young were emerging. All distances were measured in the field.*

Home burrow	Sample	Distance (m)	
distance category	size	Mean ± S.E.	Range
2- to 8-year females, interyear	24	17.4 ± 3.2	0.0– 60.0
2- to 5-year males, interyear	10	175.0 ± 25.4*	56.0– 288.0
Females' mating site(s)—her burrow that year (13 different females)	19	36.4 ± 18.1	13.6– 148.9
Males' mating site(s)—his burrow that year (5 different males)	10	176.3 ± 37.1*	106.7– 380.0
1-year females' burrow—their natal burrow	27	38.4 ± 6.3	5.5– 140.8
1-year males' burrow—their natal burrow	13	223.7 ± 39.9*	58.3– 510.0
2-year females' burrow—their natal burrow	9	47.1 ± 13.7	7.6– 132.4
2-year males' burrow—their natal burrow	7	449.7 ± 161.3*	113.0–1385.0
1-year sisters' burrows	17	38.5 ± 7.2	2.9– 115.0
1-year brothers' burrows	6	273.2 ± 49.0*	108.9– 437.8
2-year sisters' burrows	7	71.8 ± 21.2	14.0– 171.5
2-year brothers' burrows	4	325.0 ± 94.8†	87.9– 393.0
Mother—1-year daughter	21	49.7 ± 5.9	2.7– 158.0
Mother—1-year son	10	239.4 ± 37.8*	61.5– 537.6

*Differences significant, $P < .005$.

†Difference significant, $P < .01$, Mann-Whitney U test.

TABLE 2. Kinship and asymmetries in tendencies to give alarm calls among female Belding's ground squirrels. Expected calling frequencies were computed as in Fig. 3; N is the number of times ground squirrels in each category were present when a predatory mammal appeared.

Category of females	N	Number observed to call	Number expected to call if 'random'	G*	P
Reproductive,† with no known living relatives	19	14	9	5.80	<.025
Nonreproductive, with no known living relatives	14	2	7		
Reproductive, with a living daughter or granddaughter, but no other living relatives	27	18	12	5.58	<.025
Reproductive, with no known living relatives	24	5	11		
Reproductive, with their mother or at least one sister alive, but no living descendants	18	13	8	5.37	<.025
Reproductive, with no known living relatives	17	3	8		
Reproductive residents: known to have lived in the same area the previous year or years	168	64	56	4.90	<.05
Reproductive nonresidents: temporary invaders to an area (see text)	49	9	17		
Reproductive, with either their mother, a sister, or a descendant alive and present when a predatory animal appears	21	9	9	—	N.S.+
Reproductive, with at least one relative alive but not present when a predatory mammal appears	11	6	6		
Reproductive, without their mother or any sisters, but with nursing young known to be alive	46	21	22	0.38	N.S.
Reproductive, without their mother or any sisters, and whose young were destroyed	16	4	3		

*G statistic, corrected for continuity, and level of significance are given.   †'Reproductive' means pregnant, lactating, or living with postweaning young of the year.   +Not significant.

11.7 m from their natal burrows (N = 19 copulations; some females mate more than once[36]); the sisters' matings took place 39.2 ± 9.2 m from each other (N = 7 pairs of copulations by sisters).

Some male Belding's ground squirrels are apparently highly polygynous. In 1975, for example, the three most successful males in one area of Tioga Pass Meadow that was under nearly continuous observation (21 percent of the sexually active males present) accounted for 21 of 37 completed copulations (57 percent); the most successful male mated with eight different females and he accounted for 22 percent of all completed copulations. Similarly, in 1976, of ten males the top two (20 percent) accounted for 19 of 32 completed copulations (59 percent); the most successful male mated with nine different females and he accounted for 31 percent of all completed copulations[36]. Unlike males in harem-polygynous sciurid species[27,37,38], male Belding's ground squirrels do not defend mating areas or territories after mating, identifiable physical resources valuable to females or to young, or sexually receptive females. Nor do males appear to behave parentally toward their mates' offspring.

During their 4- to 6-hour period of sexual receptivity, females mate with a mean of 2.1 ± 0.2 different males (± S.E., N = 34 females, 69 copulations). Females rear their young alone, and they protect their offspring from conspecifics that find neonatal ground squirrels acceptable prey by excluding non-descendants from the area surrounding their nest burrows[36]. About the time that their mates' young are born, the males that copulated most frequently abandon areas where their mates will rear young and inhabit burrows elsewhere (Table 1); unsuccessful males do not move. The successful males usually remain near their new burrows until after they have attempted to mate there the following spring. During the lactation period, a male that had mated to completion with more than one female returned to and entered the area defended by one of his mates only once every 19.3 ± 3.2 hours (data from 7 males, 17 females); similarly, nonmates entered a female's defended area during the same period only once every 16.9 ± 4.1 hours (data from 11 females, 13 adult males). A returning mate was chased away by the resident female 42 of 53 times (79 percent). Similarly, during the lactation period, males who had either not mated at all or else had not copulated with particular females were chased, if they trespassed, from the defended areas of those nonmates 32 of 38 times (84 percent).

KINSHIP AND ASYMMETRIES IN TENDENCIES TO GIVE ALARM CALLS

When a predatory mammal appears, adult and 1-year-old female Belding's ground squirrels give alarm calls more frequently than would be expected if the animals called in direct proportion to the number of times they were present when a predator arrived (that is, expected if calls were 'random'); by contrast, males call considerably less often than would be expected under randomness (Fig. 3). Twenty-two times only males were present (that is, no females were there) when a predatory mammal appeared, and four times (18 percent) alarm calls were given by one of them. Conversely, only females were present 47 times when a predator appeared, and alarm calls were given in 40 (85 percent) of these cases. (For this comparison, the number of males present in alarm-call-evoking situations when no females were there and the number of females present when no males were there did not differ significantly; $P > .09$, Mann-Whitney U test.) Because of the matrilineal kin group structure of Belding's ground squirrel populations (Table 1) and because females are the more parental sex in this species, the sexual dimorphism in calling frequency (Fig. 3) suggests that the alarm call under consideration might function to warn kin (that is, hypothesis 3).

In apparent support of the nepotism hypothesis[2,11] are data (Table 2) suggesting that when a predatory mammal appears (i) reproductive females without living mothers, sisters, or descendants call more frequently than do non-reproductive females similarly lacking close female relatives, (ii) reproductive females without living mothers or sisters but with at least one living female descendant (that is, a daughter or a granddaughter) call more frequently than do reproductive females without living mothers, sisters, or descendants, (iii) reproductive females without living female descendants but whose mothers or at least one sister are alive call more frequently than do reproductive females lacking all three classes of close female relatives, and (iv) temporary 'invaders,' reproductive but nonresident females, known not to have lived on a study plot within Tioga Pass Meadow in the previous year or years and present less than 1 hour, call less frequently than do reproductive residents[39] (for this latter comparison, all reproductive females were considered whether or not their family members were alive).

Although the data are sparse, it appears that females with living female relatives call whether or not those family members are actually present when a predatory mammal appears (Table 2). Destruction of the current year's litter also does not seem to affect calling tendencies (Table 2).

Analysis of variance of 1974–75 data from 87 encounters between ground squirrels and predators (involving 174 different reproductive females of known age) indicates that time of year[40] has no effect on calling frequency ($F = 2.03$, d.f. $= 2$, $P = .17$), but that the age of the female does have a significant effect ($F = 19.8$, d.f. $= 1$, $P = .005$); the likelihood that alarm calls will be given by females increases with increasing age[41]. Among males, alarm calling and copulatory success seem to be unrelated. When predatory mammals appeared in 1975, seven males that had copulated at least once called no more frequently (that is, no greater percentage of the times when a predator appeared) than did eight males that had not copulated in 1975 ($P > .2$, Mann-Whitney U test). Among the seven 1975 males that copulated at least once, there was no correlation between the number of matings with different females and the percentage of alarm-call-evoking situations in which each male called ($P > .3$, Kendall's rank correlation test).

Neither the first ground squirrel that behaved as if it saw a predator (Table 3), the animal closest to the danger, nor the one closest to its own burrow always sounded the first alarm. On 54 occasions, the animal first reacting to a predator was identified and its sex was ascertained. In 6 of

TABLE 3. Behavior of Belding's ground squirrels toward pedatory mammals or toward alarm calls from conspecifics. Data are from 102 ground squirrel-predator interactions (1974 to 1976).

Class of animal	None	Sits up but does not change location	Observed responses					
			Runs toward					
			Con-specific	Rock	Bush	Mouth of any burrow other than home	Mouth of the home burrow	Defended area
Ground squirrels within the defended area surrounding the burrow								
Adult females (2 to 8 years)	6 (8%)	29 (37%)	3 (4%)	33 (43%)	1 (1%)	2 (3%)	3 (4%)	—
1-year females	2 (5%)	14 (37%)	2 (5%)	15 (40%)	0 (0%)	2 (5%)	3 (8%)	—
Juveniles of both sexes	2 (5%)	13 (33%)	3 (8%)	6 (15%)	1 (3%)	6 (15%)	8 (21%)	—
Total	10 (7%)	56 (36%)	8 (5%)	54 (35%)	2 (1%)	10 (7%)	14 (9%)	—
Ground squirrels not within the defended area surrounding the burrow								
Adult females	2 (3%)	12 (23%)	3 (5%)	15 (29%)	1 (2%)	1 (2%)	3 (5%)	16 (31%)
1-year females	1 (4%)	5 (22%)	0 (0%)	8 (35%)	0 (0%)	1 (4%)	2 (8%)	6 (26%)
Juveniles of both sexes	0 (0%)	2 (11%)	1 (6%)	6 (33%)	0 (0%)	4 (22%)	1 (6%)	4 (22%)
Total	3 (3%)	19 (20%)	4 (4%)	29 (31%)	1 (1%)	6 (6%)	6 (6%)	26 (29%)
Adult males (2 to 5 years)	5 (11%)	16 (35%)	4 (9%)	17 (37%)	1 (2%)	1 (2%)	2 (4%)	*
1-year males	3 (11%)	9 (32%)	2 (7%)	13 (46%)	0 (0%)	0 (0%)	1 (4%)	*
Total	8 (11%)	25 (34%)	6 (8%)	30 (41%)	1 (1%)	1 (1%)	3 (4%)	*

*Males do not defend areas surrounding burrows as do females.

the 31 times that an adult male reacted first (19 percent), the first-reacting male also called first, and in 9 of the 23 times that a reproductive female reacted first (39 percent), the first-reacting female also called first. In 68 instances, the ground squirrel closest to a predator when the predator was first seen by a human observer was identified and its sex was ascertained. In 5 of the 36 times that an adult male was the closest (14 percent), the closest male also called first, and in 9 of the 32 times that a reproductive female was the closest (28 percent), the closest female also called first. Among reproductive residents, 21 females giving alarm calls were no closer to their home burrows than were 19 simultaneously present noncallers ($P > .1$, Mann-Whitney U test). Thus, when a predatory mammal appears, old (that is, 4 to 7+ years), reproductive, resident females with living kin are most likely to call, while males are the most consistent noncallers. Again the implication is that warning family members, hypothesis 3, is a likely function of this alarm call.

DISCRIMINATING AMONG THE ALTERNATIVE HYPOTHESES

Could these data be better explained by any of the five hypotheses alternative to nepotism? Contrary to hypothesis 1, alarm calls did not divert predators' attention to other prey by causing pandemonium among the ground squirrels, and the animals did not aggregate to mob or to flee from predators (Table 3). Four times an adult female chased a long-tailed weasel from the neighborhood of her burrow, and in none of these cases did any conspecifics aid her[42]. Whether or not they were near their burrows, most ground squirrels either sat up or ran to a rock upon sighting a predatory mammal or upon hearing an alarm call (Table 3). Occasionally juveniles squeaked when hand-held, and these screams from captured individuals sometimes attracted their mothers or other reproductive females. Such squeaks were clearly different from the alarm calls under discussion (that is, Fig. 2), and they ceased 3 to 4 weeks after juveniles appeared above the ground for the first time. First callers and other alarmers did not seek cover in the center of an aggregation of conspecifics. Neither did alarm callers appear to sequester information on the whereabouts of approaching predators, and the calls did not seem ventriloquial to us or, apparently, to predators (below). Alarm callers usually sat upright, often on prominent rocks, and looked directly toward the advancing predator, thereby seemingly directing the attention of conspecifics toward it[43]. Indeed, I could often locate the predator by following the gaze of several alerted animals, whether or not they were calling. I do not know whether ground squirrels also use this cue. However, in 11 instances a ground squirrel probably cold not see an advancing predator because of the ground squirrel's position in a swale; on eight of these occasions (73 percent), the ground squirrel sat up and oriented itself in the same direction as a conspicuous, calling conspecific, thus toward the apparently unseen predator. Only one of nine times (11 percent) did a ground squirrel in the same swale orient toward an apparently unseen predator when no conspecific was calling. Thus, no evidence supports the hypothesis that the alarm call results in the diversion of predators' attention to other prey (that is, hypothesis 1).

Members of all five mammalian predator species appeared undeterred by ground squirrel alarm calls, suggesting that the call does not function to discourage predator pursuit (that is, hypothesis 2). Indeed, members of all five species stalked or chased alarm callers, suggesting that calling may in fact make alarmers more conspicuous. Three of six adult ground squirrels preyed upon during this study had called just prior to being attacked. Also, calling ground squirrels were stalked or chased by predators significantly more often than were noncallers. A marked ground squirrel was stalked or chased 22 times; 14 of 107 calling animals (13 percent) were so attacked, but only 8 of 168 noncallers (5 percent) were similarly attacked ($P < .025$, two-tailed G statistic, corrected for continuity). To test hypothesis 2 further, I considered the responses of coyotes to callers separately. Because coyotes sometimes hunted by remaining motionless or hidden near bushes for long periods as if the element of surprise were important to their success, and provided that the alarm call under consideration discourages predator pursuit by indicating that the advantage of surprise has been removed, coyotes in particular might be deterred by 'it'. A coyote caught a mountain vole (*Microtus montanus*) and behaved as if it were continuing to hunt this species or other prey on ten occasions; in these cases, 39 ground squirrels gave alarm calls and 41 were silent. Five of the 39 callers (13 percent) were apparently stalked or were chased by the predator, while only 3 of the 41 noncallers (7 percent) were similarly pursued (this difference is not significant at the $P < .05$ level, G statistic). Thus, coyotes do not turn away from calling ground squirrels; if anything they, like other predators, are attracted to callers. None of the predators seemed to be startled or confused by alarm calls. On the four occasions when we observed the behavior of a predatory mammal toward the ground squirrel that it had just killed, the predator consumed its victim, suggesting that Belding's ground squirrels are not distasteful (nor poisonous) and that, therefore, alarm calling is not an aposematic display. The abundance of noncallers and the male-bias among them (Fig. 3) do not support the second hypothesis, the lack of correspondence between the nearest ground squirrel to the predator (that is, the one likely to be in greatest proximate danger) and the first alarm caller, or the first one behaving as if it saw the predator (Table 3) and the first alarm caller also do not support the hypothesis that the alarm call functions to discourage predator pursuit (hypothesis 2).

Although this population of ground squirrels was not divided up into identifiable, physically isolated demes[13], females successfully raising young were relatively sedentary during 1974–76 (Table 1). Behaviors observed among these stable aggregations might have spread by a process of between-group selection (that is hypothesis 4; see 14, 15). Because these aggregations are composed mainly of close relatives—mothers, daughters, sisters, cousins, and nieces—the 'groups' are appropriately characterized as matrilineal kinship associations. The likelihood that female family members are consistently alerted by alarm calls and the apparent interdependence of kinship and calling (Table 2) make it impossible in this species to support between-group selection over kin selection (that is, hypothesis 3) (16, 17; but see 18). With a dog, I visited six Sierra Nevada populations of Belding's ground squirrels other than the primary population under study; all visited populations were greater than 0.5 km but less than 23 km from Tioga Pass Meadow. At least one alarm call, usually many, was heard at each soon after the dog was released. Thus I have no evidence that noncalling groups or populations of ground squirrels occur in the vicinity of Tioga Pass Meadow. These data are obviously inadequte to determine whether there are between-group or between-population differences in the percentages of alarm callers. Because I found no non-calling populations of Belding's ground squirrels and because aggregations of related females do not predictably break up, emigrate from their natal area, and reassemble with alarm callers not sharing common ancestry, however, the most important prerequisites[13-15] for the operation of between-group selection (that is, hypothesis 4) are seemingly absent.

Because female ground squirrels are more sedentary than are males (Table 1), females might be more frequently in jeopardy than males if predators return to hunt near sites of previous successful kills. Females also give alarm calls more frequently than do males (Fig. 3). Taken together, these observations suggest that the alarm call might function to reduce the likelihood of later attacks by the same predator (that is, hypothesis 5). However, mammalian predators at Tioga Pass Meadow do not preferentially return to sites of previous successes. For seven diurnal predations by coyotes and long-tailed weasels, the time between visits by a member of the successful species to a ground squirrel's defended area contiguous to one on which a kill had been made, 20.9 ± 6.2 days, was not different from ($P \geq .10$, Mann-Whitney U tests) the time between visits to seven randomly chosen defended areas, 18.9 ± 8.4 days, on which ground squirrels had never been captured [this comparison was made five times with seven different, randomly chosen defended areas each time; in no case were any significant differences found]. If predators did return to hunt near sites of previous successes, under hypothesis 5 young females should give alarm calls more frequently than older females; because the probability of dying increases with increasing female age in this species[36], young females would be in jeopardy from returning predators more often in their lifetimes that would older females (but see 41). Contrary to the prediction of decreases in calling with increases in female age, tendencies to give alarm calls increase with increasing female age. Discrimination among alarm-call-evoking situations, apparently on the basis of kinship with individuals likely to be alerted (Table 2), is also not predicted by hypothesis 5, but this observed discrimination does support the hypothesis that one function of the alarm call is to warn relatives (that is, hypothesis 3).

Because aggregations of (closely related) female Belding's ground squirrels are more stable through time than are male-male or male-female associations (Table 1), reciprocity[20] might be more likely to occur among females than among males. Therefore, the sexual dimorphism in probability of giving an alarm call (Fig. 3) could indicate that the call functions to warn conspecifics likely to reciprocate (that is, hypothesis 6). If so, the 'reciprocators' are also family members, and reciprocation might therefore benefit callers genotypically as well as phenotypically[21]. Because reciprocity, as Trivers[20] formulated the hypothesis, refers only to an exchange of phenotypic benefits, circumstances (20, p. 35) '. . . when the recipient is so distantly related to the organism performing the altruistic act that kin selection can be ruled out,' the alarm call under discussion does not function only in the context described by hypothesis 6. The degree to which alarm callers discriminate against distantly related or unrelated individuals known not to call might indicate the degree to which the alarm call functions to warn phenotypic reciprocators[20,22]. Limited evidence suggests that the presence of certain kinds of noncallers at least does not deter females with living relatives from calling. Using data from 28 encounters between predatory mammals and reproductive females whose mothers or at least one sister or daughter were alive, I compared the time between the moment a human observer first saw a predator and the first alarm call and the percentage of callers versus noncallers under two circumstances: when no noncallers were present, and when at least one unrelated male, temporary female 'invader,' or one nonreproductive female not known to be related to any of the residents in a study plot was present. In neither of these comparisons did callers' responses differ significantly on the basis of the presence of noncallers ($P \geq .2$ for each comparison, Mann-Whitney U tests). In assessing the importance of this apparent lack of a difference, note that discrimination on the basis of whether certain relatives are alive does occur (Table 2). In other words, females call more frequently when relatives might be alerted; they refrain from calling when no kin are alive despite being surrounded by (unrelated) females, members of the sex that calls. Although reciprocation might occur between related ground squirrels with reciprocators benefiting genotypically as well as phenotypically[15]—because nonreciprocators are not obviously discriminated against when rather subtle discrimination on the basis of relatedness apparently occurs—it is not possible to support the phenotypic reciprocity hypothesis (that is, hypothesis 6)[20] apart from the nepotism hypothesis (that is, hypothesis 3).

Conclusions

My observations suggest that it is possible to begin discriminating among theoretical alternative functions of alarm calls and other behaviors that, because they may be phenotypically hazardous, appear altruistic. Data and arguments deriving from them imply that, of the six hypothesized alternative benefits of giving alarm calls, warning relatives, hypothesis 3 is a likely function of the alarm call that Belding's ground squirrels give when terrestrial predators approach. Regarding the other possible functions of this alarm call, no evidence supports hypotheses 1 (diverting predators' attention), 2 (discouraging predator pursuit) or 5 (reducing the likelihood of later attacks by the same predator). That the alarm call may function to help the group (hypothesis 4) or to warn reciprocators (hypothesis 6) is possible; but when assumptions of the fourth and sixth hypotheses and predictions derived from them and from the hypothesis 3 that the call alerts relatives are contrasted and are compared with field observations of the ground squirrels' behavior, both appear to be at most less important functions than warning kin.

Among the sciurids in which males give little or no parental care and in which matrilineal kin groups are known or are appropriately suspected to be a basic population unit[34], there exist similarities in the form[29-31] and female sex- and age-specificity of alarm calls to terrestrial predators[11,29,34,44]. Further, in at least one sciurid in which males have harems and live with and probably protect their mates and their mates' offspring year-round, harem-males call most frequently[45]. These observations suggest that warning kin might be a common function of sciurid alarm calls to predatory mammals and they imply that asymmetries in tendencies to call may be expressions of discriminative nepotism[21].

1. For example, Barash, D. P. [*Am. Midl. Nat.* **94**, 468 (1975)] reported that 'During 7 years of study of free-living marmots (Rodentia: Sciuridae), I observed eight instances of predation and numerous cases of alarm-calling in 3017 hr. of direct field observations on five marmot species . . . .'

2. Hamilton, W. D. *J. Theor. Biol.* **7**, 1 (1964); West Eberhard, M. J. *Q. Rev. Biol.* **50**, 1 (1975).

3. Hamilton, W. D. *J. Theor. Biol.* **31**, 295 (1971); Vine, I. *ibid.* **30**, 405 (1971); Treisman, M. *Anim. Behav.* **23**, 779 (1975).

4. As suggested by Rohwer, S., Fretwell, S. D. & Tuckfield, R. C. [*Am. Midl. Nat.* **96**, 418 (1976)].

5. For example, Hoogland, J. L. & Sherman, P. W. *Ecol. Monogr.* **46**, 33 (1976); Windsor, D., Emlen, S. T. *Condor* **77**, 359 (1975).

6. Charnov, E. L., Krebs, J. R. *Am. Nat.* **109**, 107 (1975).

7. Owens, N. W., & Goss-Custard, J. D. *Evolution* **30**, 397 (1976).

8. Perrins, C. *Ibis* **110**, 200 (1968).

9. Smythe, N. *Am. Nat.* **104**, 491 (1970); Zahavi, A. in *Evolutionary Aspects of Ecology*, Stonehouse, B., Perrins, C. M. Eds. (Macmillan, London, in press).

10. Driver, P. M., & Humphries, D. A. *Ibis* **111**, 243 (1969); Humphries, D. A. & Driver, P. M. *Science* **156**, 1767 (1967).

11. Maynard Smith, J. *Am. Nat.* **99**, 59 (1965); Dunford, C. *Am. Nat.* **111**, 782 (1977).

12. Williams, G. C. *Adaptation and Natural Selection* (Princeton Univ. Press, Princeton, N.J., 1966). See also Power, H. W. *Science* **189**, 142 (1975); Hirth, D. H. & McCullough, D. R. *Am. Nat.* **111**, 31 (1977).

13. Levins, R. *Am. Math. Soc. Publ.* **2**, 77 (1970); Levin, B. R. & Kilmer, W. L. *Evolution* **28**, 527 (1974); Eshel, I. *Theor. Popul. Biol.* **3**, 258 (1972).

14. Wilson, D. S. *Proc. Natn. Acad. Sci. U.S.A.* **72**, 143 (1975); *Am. Nat.* **111**, 157 (1977).

15. A similar suggestion was made by Hamilton,

W. D. [in *Biosocial Anthropology*, Fox, R. Ed. (Wiley, New York, 1975), p. 133].

16. Maynard Smith, J. *Q. Rev. Biol.* **51**, 277 (1976).

17. West Eberhard, M. J. *ibid.*, p. 89; Maynard Smith, J. *Nature (London)* **201**, 1145 (1964); Alexander, R. D. & Borgia, D. in preparation; Brown, J. L. *Nature (London)* **211**, 870 (1966). See also Gadgil, M. *Proc. Natl. Acad. Sci. U.S.A.* **72**, 1199 (1975).

18. Wilson, E. O. *BioScience* **23**, 631 (1973).

19. Curio, E. *The Ethology of Predation*, vol. 7 of the series *Zoophysiology and Ecology*, Farner, D. S., Ed. (Springer-Verlag, Berlin, 1976), p. 58.

20. Trivers, R. L. *Q. Rev. Biol.* **46**, 35 (1971).

21. Alexander, R. D. *Annu. Rev. Ecol. Syst.* **5**, 325 (1974).

22. Olive baboons, *Papio anubis*: see Packer, C. *Nature (London)* **265**, 441 (1977).

23. The study area, an alpine meadow about 1000 m long by 450 m wide, adjoins and lies just east of California State Highway 120, between Lake Tioga and the Yosemite National Park boundary at Tioga Pass, Mono County. For a description of the area and the annual cycle of ground squirrels there, see Morton, M. L. *Bull. South. Calif. Acad. Sci.* **74**, 128 (1975); ——, Maxwell, C. S., Wade, C. E. *Great Basin Nat.* **34**, 121 (1974); Morton, M. L. & Gallup, J. S. *ibid.* **35**, 427 (1975).

24. Hall, E. R. & Nelson, K. R. *The Mammals of North America* (Ronald, New York, 1959), vol. 1, p. 340; Durrant, S. D. & Hanson, R. M. *Syst. Zool.* **3**, 82 (1954).

25. Turner, L. W. thesis, University of Arizona, Tucson (1972).

26. Hoogland, J. L. thesis, University of Michigan, Ann Arbor (1977); in preparation.

27. King, J. A. *Contrib. Lab. Vertebr. Biol. Univ. Mich.* **67**, 1 (1955).

28. Koford, C. B. *Wildl. Monogr.* **3**, 1 (1958).

29. For a discussion of the genus *Spermophilus*, see Manville, R. H. *J. Mammal.* **40**, 26 (1959);

Balph, D. M. & Balph, D. F. *ibid.* **47**, 440 (1966); Harris, J. P. W. thesis, University of Michigan, Ann Arbor (1967); Betts, B. J. *Anim. Behav.* **24**, 652 (1976); Owings, D. H., Borchert, M., Virginia, R. *ibid.* **25**, 221 (1977). For a discussion of the genus *Eutamias*, see Brand, L. R. *ibid.* **24**, 319 (1976). For a discussion of the genus *Tamias*, see Dunford, C. *Behavior* **36**, 215 (1970). Species in the genus *Cynomys* give mono- or bisyllable alarm barks in the 1- to 7-khz range, several in succession. In this regard see Waring, G. H. *Am. Midl. Nat.* **83**, 167 (1970); Pizzimenti, J. J. & McClenaghan Jr., L. R. *ibid.* **92**, 130 (1974).

30. For example, for Arctic ground squirrels, *Spermophilus undulatus*, see Melchior, H. R. *Oecologia (Berlin)* **7**, 184 (1971).

31. Turner, L. W. *J. Mammal.* **54**, 990 (1973).

32. For descriptions of audibly distinct alarm calls to aerial and terrestrial predators in California ground squirrels, *Spermophilus beecheyi*, see Linsdale, J. M. *The California Ground Squirrel* (Univ. of California Press, Berkeley, 1946); Fitch, H. S. *Am. Midl. Nat.* **39**, 513 (1948); Owings, D. H. & Virginia, R. A. Z. *Tierpsychol.*, in press.

33. Marler, P. *Naure (London)* **176**, 6 (1955); *Ibis* **98**, 231 (1956); *Behavior* **11**, 13 (1957); Armstrong, E. A. *A Study of Bird Song* (Oxford Univ. Press, London, 1963).

34. For a discussion of *Spermophilus armatus*, see Walker, R. E. thesis, Utah State University, Logan (1968); Slade, N. A. & Balph, D. F. *Ecology* **55**, 989 (1974). For a discussion of *S. richardsoni*, see Yeaton, R. I. *J. Mammal.* **53**, 139 (1972); Michener, D. R. thesis, University of Saskatchewan, Regina (1972); ___ & Michener, G. R. *J. Mammal.* **52**, 853 (1971); *Ecology* **54**, 1138 (1973). For a discussion of suggestive data on *S. tridecemlineatus*, see Rongstad, O. J. *J. Mammal.* **46**, 76 (1965); McCarley, H. *ibid.* **47**, 294 (1966).

35. For a discussion of *Marmota flaviventris*, see Armitage, K. B. & Downhower, J. F. *Ecology* **55**, 1233 (1974).

36. Sherman, P. W. thesis, University of Michigan, Ann Arbor (1976).

37. For a discussion of *Marmota flaviventris*, see Downhower, J. F. & Armitage, K. B. *Am. Nat.* **105**, 355 (1971); Armitage, K. B. *J. Zool.* **172**, 233 (1974). For a discussion of *M. calligata*, see Barash, D. P. *J. Mammal.* **56**, 613 (1975); of *M. olympus*, see ___, *Anim. Behav. Monogr.* **6**, 171 (1973); of *Spermophilus columbianus*, see Steiner, A. L. *Rev. Comp. Anim.* **4**, 23 (1970).

38. For reports of the defense of sexually receptive females by males among sciurids that are probably not harem-polygynous, see: for *Spermophilus undulatus*, Carl, E. A. *Ecology* **52**, 395 (1971); for *Sciurus aberti*, Farentinos, R. C. *Anim. Behav.* **20**, 316 (1972).

39. Similarly, Carl, E. A. [see (38)] noted that for a nonreproductive, transient or 'refugee' population of *Spermophilus undulatus*, that included members of both sexes, 'The population was singularly silent; only occasionally did I hear a squirrel vocalization, in sharp contrast to the barrage of alarm calls that greeted me whenever I walked across the study area' (*38*, p. 410).

40. The 1974 and 1975 breeding seasons were divided into three segments for each reproductive female considered in the analysis: (i) spring emergence-parturition, (ii) parturition-first appearance of young above ground, and (iii) first appearance of young-fall disappearance of the female (hibernation). The arc sine square root transformation was used to produce normality in the data analyzed. The data consist of the percentage of times that calls were given by females of each age class, 1 to 7+ years, when a predatory mammal appeared.

41. Barash, D. P. (1) observed an apparently similar effect of age on tendency to give alarm calls among marmots. The effect that he and I observed might result if older, more experienced females either (i) are more familiar with routes of escape near their burrows, thus more able than less experienced females to evade predators once they have rendered themselves conspicuous by giving an alarm call, or (ii) are redirecting nepotism from current or future (expected) young to offspring or grandchildren that have reached reproductive age, or both (i) and (ii). Advantages of assisting descendants whose likelihood of future reproduction is higher than a female's own reproductive potential may favor increasing nepotism with advancing age among female mammals generally [a similar suggestion was made by Hrdy, S. B. & Hrdy, D. B. *Science* **193**, 913 (1976)]. Menopause-like termination of reproduction coupled with extensive maternal care in, for example, elephants [Laws, R. M., Parker, I. S. C., Johnstone, R. C. B. *Elephants and Their Habits* (Clarendon, Oxford, 1975)] may suggest that nepotism is sometimes completely redirected toward relatives that have survived to reproductive age [see also (21)].

42. Turner, L. W. (25, 31) made similar observations on *Spermophilus beldingi*. Warren, E. R. [*J. Mammal.* **5**, 265 (1924)] also reported single *S. armatus* chasing long-tailed weasels while conspecific ground squirrels looked on. Once Loehr, K. [thesis, University of Nevada, Reno (1974), p. 22] observed '. . . several adult Belding Ground Squirrels . . .' chasing a weasel.

43. Melchior, H. R. (30) and Carl, E. A. (*38*) reported that *Spermophilus undulatus* behave similarly. Farentinos, R. C. [*Z. Tierpsychol.* **34**, 441 (1974)] reports parallel observations for *Sciurus aberti*.

44. See also, Fitzgerald, J. P. & Lechleitner, R. R. *Am. Midl. Nat.* **92**, 146 (1974); Grizzell Jr., R. A. *ibid.* **53**, 257 (1955).

45. For a discussion of *Marmota marmota*, see Barash, D. P. *Anim. Behav.* **24**, 27 (1976).

46. I thank my field assistants Blumer, L., Dunny, K., Flinn, S., Flinn, M., Kagarise, C., Kuchapsky, D., Mulder, B., Odenheimer, J., Roth, M. & Schultz, B. The support of Morton, M. L. was invaluable. For other assistance, I thank Alexander, R., Blick, J.,

Hoogland, J., Huey, R., Kagarise, C., Koford, R., Payne, R., Pitelka, F., Stearns, S. & Tinkle, D. The Southern California Edison Company provided housing, and the Clairol Company donated hair dye. Supported by NSF grant GB-43851, the Theodore Roosevelt Memorial Fund, the Museum of Zoology and the Rackham School of Graduate Studies at the University of Michigan, and the Museum of Vertebrate Zoology and the Miller Institute at the University of California, Berkeley.

Paul W. Sherman is in the Department of Zoology and the Museum of Vertebrate Zoology at the University of California, Berkeley.

Reciprocal altruism in Papio anubis

C. PACKER

ALTRUISM is behaviour that benefits another individual at some cost to the altruist, costs and benefits being measured in terms of individual fitness. 'Reciprocal altruism' (ref. 1) implies the exchange of altruistic acts between unrelated individuals as well as between relatives. If the benefits to the recipient of an altruistic act exceed the costs to the altruist, and if the recipient is likely to reciprocate at a later time, then the cumulative benefits for both individuals will have exceeded the cumulative costs of their altruism. Natural selection would favour individuals that engaged in reciprocal altruism if they distributed their altruism with respect to the altruistic tendencies of the recipient, preferring individuals that were most likely to reciprocate and excluding nonaltruists from the benefits of further altruism. This model has been difficult to test because it is usually impossible to be certain that an example of altruism is not the product of 'kin selection'[2]. The genetic relationships between individuals in animal populations are seldom known and reciprocal altruism can only be cited when it can be found to occur regularly between unrelated individuals. I report here that altruistic behaviour involving the formation of coalitions among male olive baboons (*Papio anubis*) fulfils the criteria for reciprocal altruism.

Eighteen adult male *P. anubis* in three troops at Gombe National Park, Tanzania were studied for more than 1,100 h, between May and December 1972 and from June 1974 to May 1975. All data were collected on a focal sample basis[3]. Focal individuals are referred to as 'targets'. Each animal was observed regularly for a fixed period. Observations were made on foot at 5–10 m from the animal. Baboon studies have been in progress since 1967 and data concerning blood relationships and dates of transfer between troops are available. All males leave their natal troop at Gombe, and the males who breed within a particular troop are those that have transferred into that troop from elsewhere[4]. Males known or thought to have transferred into their troop of residence are termed 'adult males'. Sexually mature males still residing in their natal troop are 'natal males'. During the study, there were three, six to eleven, and eight to eleven adult males in the respective study troops.

The effectiveness of temporary coalitions of adult male *Papio* spp. during aggressive interactions against a single opponent has been described in *P. anubis*[5] and in *P. ursinus*[6]. Encounters between coalitions and single opponents in *P. cynocephalus* did not seem to affect subsequent dyadic encounters between a coalition member and the opponent[7]. Ransom noted that coalitions of *P. anubis* at Gombe were generally formed by one male enlisting a partner to help fight against an opponent[8]. The partner had not been involved directly in the encounter with the opponent before his enlistment. Coalitions were sometimes formed in attempts to separate an opponent from an oestrous female. *P. anubis* form exclusive consort pairs lasting for up to several days. If the female became available after such an attempt, she could be taken by only one of the two coalition partners.

Attempts at enlisting a coalition partner are referred to as 'soliciting'—a triadic interaction in which one individual, the enlisting animal, repeatedly and rapidly turns his head from a second individual, the solicited animal, towards a third individual (opponent), while continuously threatening the third. The function of headturning by the enlisting animal is to incite the solicited animal into joining him in threatening the opponent. An 'occasion' of soliciting is a bout of the behaviour followed by a gap of more than 10 s. The distribution of the number of 'occasions' of soliciting to the same partner per observation period was tested against a cumulative binomial distribution and did not differ significantly from expected values. Therefore, 'occasions' of soliciting are considered to be statistically independent from each other. For every occasion that soliciting involved the target male during an observation period, the identity and actions (enlisting, solicited, or opponent) of each participant were recorded. 'Occasions' were not recorded consistently in 1972, so data from that period are not included in measures of frequency. Although coalitions occasionally formed spontaneously, only those that resulted from soliciting are considered.

There were 140 examples of one adult male soliciting another during the 1974–75 study period. Soliciting resulted in coalitions on 97 occasions. On 20 occasions the opponent was consorting with an oestrous female and adult males were more likely to join a coalition if the opponent was in consort ($2 \times 2 \chi^2 = 5.91$, $P < 0.02$, $n = 140$). On six occasions during both study periods the formation of a coalition directed against a consorting male resulted in the loss of the female by the opponent. In all six cases the female ended up with the enlisting male of the coalition ($P = 0.032$, two-tailed, sign test); the solicited male generally continued to fight the opponent while the enlisting male took over the female. In each of these cases the solicited male risked injury from fighting the opponent while the enlisting male gained access to an oestrous female. In most other examples of soliciting, no resource appeared to be at stake. The greater willingness to join a coalition against a consorting male may be related to the greater benefits that the altruism bestows on the recipient in those cases.

Thirteen different pairs of males reciprocated in joining coalitions at each other's request on separate occasions. These pairs comprised 12 different males. In six of these pairs, both pair members successfully enlisted their partner against the same opponent. Individual males which most frequently gave aid were those which most frequently received aid.

There was a strong correlation between the frequency with which adult males joined coalitions and the frequency with which they successfully enlisted coalition partners ($\hat{\rho} = 0.84$, $z = 2.87$, $P < 0.004$, Kraemer test[9] based on Spearman correlations).

Each male tended to request aid from an individual who in turn requested aid from him. Using only one occasion of soliciting per pair per observation period (since spatial patterns are relatively stable), the distribution of soliciting by each of the 18 target males was examined to find which other male each target solicited most often, that is, his 'favourite partner'. Only the ten targets who solicited other males on four or more occasions were included. After the favourite partner was found for each target male, the distribution of soliciting by each partner was similarly examined. For nine out of ten target males, the favourite partner in turn solicited the target male more often than the average number of occasions that the partner solicited all adult males in their troop ($P = 0.022$, two-tailed, sign test) (Table 1). These results suggest that preferences for particular partners may be partly based on reciprocation.

Table 1 Soliciting activity of favourite partners of target males

Target	No. of occasions FP solicited target	Mean no. of occasions FP solicited each male	FP solicited target more; + or less; − than average
BBB	3	2.7	+
CRS	3	1.2	+
DVD	0	1.0	−
EBN	5	1.0	+
GRN	1	0.7	+
JNH	2	1.1	+
LEO	4	2.7	+
MNT	1	0.4	+
WDY	5	2.7	+
WTH	4	2.8	+

FP, favourite partners.

It is difficult to test whether a non-altruist is excluded from further altruistic exchanges; a refusal to join a coalition in any given instance may occur because the costs to the potential altruist on that occasion are particularly high. There is evidence, however, that adult males are more likely to join coalitions with individuals which in turn would be able to help them. Females and juveniles ('non-males') solicited adult males 12 times, but adult males never solicited 'non-males' ($P < 0.001$, two-tailed, sign test). Adult males were less likely to respond to the solicitings of 'non-males' than they were to other adult males (2×2 $\chi^2 = 7.76$, $P < 0.01$, $n = 152$), even though 'non-males' solicited adult males against other 'non-males' far more often that adult males solicited other adult males against 'non-males' ($2 \times 2 \; \chi^2 = 43.76$, $P < 0.001$, $n = 110$). Adult male baboons are very much larger than either adult females or juveniles, so that even though the aid of an adult male to a 'non-male' would generally incur small costs to the adult male, the benefits to the male from having a 'non-male' partner would be trivial, since a 'non-male' could not provide effective help in an encounter against another adult male and an adult male would not need help in fighting a 'non-male'.

The adult males in the troops studied transferred into their troop of residence singly[10]. The origins of many males entering a study troop were not definitely known, so it was impossible to determine the precise degree of relatedness between most adult males. The previous troop and dates of transfer of each male were known, however, In 4 of the 13 reciprocating pairs, both partners were first observed as young adults in different troops when they probably had not yet transferred for the first time. They did not reside in the same troop for another 5 yr. One pair not included earlier comprised an adult male and a 'natal' male which were known to have been born in separate troops. Although there is no proof that any of these individuals are completely unrelated, it is unlikely that they are close relatives.

In contests between a coalition and an opponent a previously unin-volved partner may benefit the enlisting male by reducing the latter's risk of injury in fighting the opponent. (Although only one male has been known to die from wounds received in a fight since 1967, non-fatal wounds are common.) By participating in a fight in which he would not have been otherwise involved, his aid will have been at some cost to himself. Whether or not the benefits to the recipient exceed the costs to the altruist (as required by the model) is difficult to determine. But, for the opponent, the potential costs in facing two males simultaneously rather than only one might be so much greater that it would often be to his advantage to avoid the coalition without fighting. If so, then the actual costs to the altruist are less than the reduction in costs to the enlisting male since the enlisting male would then be less likely actually to fight the opponent. When the formation of a coalition involves gaining access to an oestrous female further benefits are involved. Coalitions are the most common way of aggressively taking over an oestrous female from a consorting male at Gombe (D. A. Collins, in preparation). During the reproductive life of a male, which extends over 10 yr, there may be a large number of situations where it would be advantageous to enlist a coalition partner in an encounter against a consorting male. The number of offspring that a male sired as a result of participating in reciprocating coalitions would be greater than if he did not, while his lifespan would probably not be appreciably shortened by aiding coalition partners.

There are probably occasions when altruism is not involved. For example, the solicited male may sometimes join a coalition when there is a prospect of immediate benefit to himself. In such cases, explanations of the animal's behaviour are not necessary; the behaviour is not detri-mental to his own fitness. The occurrence of genuine altruism, however, seems to be common enough to demand an alternative explanation.

I thank J. Goodall and the government of Tanzania for making this research possible, and T. H. Clutton-Brock, D. A. Collins, P. H. Harvey, H. C. Kraemer, J. Maynard Smith, A. E. Pusey, and R. W. Wrangham for advice. Field work was supported by grants from the Ford Foundation and the W.T. Grant Foundation.

1. Trivers, R. L. *Q. Rev. Biol.* **46**, 35–57 (1971).
2. Hamilton, W. D. J. *Theor. Biol.* **7**, 1–52 (1964).
3. Altmann, J. *Behaviour* **49**, 227–267 (1974).
4. Packer, C. *Nature* **255**, 219–220 (1975).
5. Hall, K. R. L. & DeVore, I. in *Primate Behaviour* (ed. by DeVore, I.), (Holt, New York, 1965).
6. Stoltz, L. P. & Saayman, G. S. *Ann. Transv. Mus.* **26**, 99–143 (1970).

7. Hausfater, G. *Contrib. Primatol.* **7** (1975).
8. Ransom, T. W. thesis, Univ. California, Berkeley (1971).
9. Kraemer, H. C. *Psychometrika* **40**, 473–484 (1975).
10. Packer, C. *Proc. Sixth Intl. Congr. Primatol.* (ed. by Harcourt, C. & Chivers, D. J.) (Academic, London, in the press).

C. Packer prepared this article while at the School of Biological Sciences, University of Sussex and is now in the Allee Laboratory of Animal Behaviour, Chicago.

This article was first published in *Nature* Vol. **265**, pp. 441–443; 1977.

The evolution of cooperation

ROBERT AXELROD AND WILLIAM D. HAMILTON

Cooperation in organisms, whether bacteria or primates, has been a difficulty for evolutionary theory since Darwin. On the assumption that interactions between pairs of individuals occur on a probabilistic basis, a model is developed based on the concept of an evolutionarily stable strategy in the context of the Prisoner's Dilemma game. Deductions from the model, and the results of a computer tournament show how cooperation based on reciprocity can get started in an asocial world, can thrive while interacting with a wide range of other strategies, and can resist invasion once fully established. Potential applications include specific aspects of territoriality, mating, and disease.

THE theory of evolution is based on the struggle for life and the survival of the fittest. Yet cooperation is common between members of the same species and even between members of different species. Before about 1960, accounts of the evolutionary process largely dismissed cooperative phenomena as not requiring special attention. This position followed from a misreading of theory that assigned most adaptation to selection at the level of populations or whole species. As a result of such misreading, cooperation was always considered adaptive. Recent reviews of the evolutionary process, however, have shown no sound basis for a pervasive group-benefit view of selection; at the level of a species or a population, the processes of selection are weak. The original individualistic emphasis of Darwin's theory is more valid[1,2].

To account for the manifest existence of cooperation and related group behavior, such as altruism and restraint in competition, evolutionary theory has recently acquired two kinds of extension. These extensions are, broadly, genetical kinship theory[3] and reciprocation theory[4,5]. Most of the recent activity, both in field work and in further developments of theory, has been on the side of kinship. Formal approaches have varied, but kinship theory has increasingly taken a gene's-eye view of natural selection[6]. A gene, in effect, looks beyond its mortal bearer to interests of the potentially immortal set of its replicas existing in other related individuals. If interactants are sufficiently closely related, altruism can benefit reproduction of the set, despite losses to the individual altruist. In accord with this theory's predictions, apart from the human species, almost all clear cases of altruism, and most observed cooperation, occur in contexts of high relatedness, usually between immediate family members. The evolution of the suicidal barbed sting of the honeybee worker could be taken as a paradigm for this line of theory[7].

Conspicuous examples of cooperation (although almost never of ultimate self-sacrifice) also occur where relatedness is low or absent. Mutualistic symbioses offer striking examples such as these: the fungus and alga that compose a lichen; the ants and ant-acacias, where the trees house and feed the ants which, in turn, protect the trees[8]; and the fig wasps and fig tree, where wasps, which are obligate parasites of fig flowers, serve as the tree's sole means of pollination and seed set[9]. Usually the course of cooperation in such symbioses is smooth, but sometimes the partners show signs of antagonism, either spontaneous or elicited by particular treatments[10]. Although kinship may be involved, as will be discussed later, symbioses mainly illustrate the other recent extension of evolutionary theory, the theory of reciprocation.

Cooperation per se has received comparatively little attention from biologists since the pioneer account of Trivers[5]; but an associated issue, concerning restraint in conflict situations, has been developed theoretically. In this connection, a new concept, that of an evolutionarily stable strategy, has been formally developed[6,11]. Cooperation in the more normal sense has remained clouded by certain difficulties, particularly those concerning initiation of cooperation from a previously asocial state[12] and its stable maintenance once established. A formal theory of cooperation is increasingly needed. The renewed emphasis on individualism had focused on the frequent ease of cheating in reciprocatory arrangements. This makes the stability of even mutualistic symbioses appear more questionable than under the old view of adaptation for species benefit. At the same time other cases that once appeared firmly in the domain of kinship theory now begin to reveal relatedness of interactants that are too low for much nepotistic altruism to be expected. This applies both to cooperative breeding in birds[13] and to cooperative acts more generally in primate groups[14]. Here either the appearances of cooperation are deceptive—they are cases of part-kin altruism and part cheating—or a larger part of the behavior is attributable to stable reciprocity. Previous accounts that already invoke reciprocity, however, underemphasize the stringency of its conditions[15].

Our contribution in this area is new in three ways.

(1) In a biological context, our model is novel in its probabilistic treatment of the possibility that two individuals may interact again. This allows us to shed new light on certain specific biological processes such as aging and territoriality.

(2) Our analysis of the evolution of cooperation considers not just the final stability of a given strategy, but also the initial viability of a strategy in an environment dominated by noncooperating individuals, as well as the robustness of a strategy in a variegated environment composed of other individuals using a variety of more or less sophisticated strategies. This allows a richer understanding of the full chronology of the evolution of cooperation that has previously been possible.

(3) Our applications include behavioral interaction at the microbial

level. This leads us to some speculative suggestions of rationales able to account for the existence of both chronic and acute phases in many diseases, and for a certain class of chromosomal nondisjunction, exemplified by Down's syndrome.

STRATEGIES IN THE PRISONER'S DILEMMA

Many of the benefits sought by living things are disproportionally available to cooperating groups. While there are considerable differences in what is meant by the terms 'benefits' and 'sought,' this statement, insofar as it is true, lays down a fundamental basis for all social life. The problem is that while an individual can benefit from mutual cooperation, each one can also do even better by exploiting the cooperative efforts of others. Over a period of time, the same individuals may interact again, allowing for complex patterns of strategic interactions. Game theory in general, and the Prisoner's Dilemma game in particular, allow a formalization of the strategic possibilities inherent in such situations.

The Prisoner's Dilemma game is an elegant embodiment of the problem of achieving mutual cooperation[16], and therefore provides the basis for our analysis. To keep the analysis tractable, we focus on the two-player version of the game, which describes situations that involve interactions between pairs of individuals. In the Prisoner's Dilemma game, two individuals can each either cooperate or defect. The payoff to a player is in terms of the effect on its fitness (survival and fecundity). No matter what the other does, the selfish choice of defection yields a higher payoff than cooperation. But if both defect, both do worse than if both had cooperated.

Figure 1 shows the payoff matrix of the Prisoner's Dilemma. If the other player cooperates, there is a choice between cooperation which yields R (the reward for mutual cooperation) or defection which yields T

	Player B	
	C Cooperation	D Defection
Player A		
C Cooperation	R=3 Reward for mutual cooperation	S=0 Sucker's payoff
D Defection	T=5 Temptation to defect	P=1 Punishment for mutual defection

FIG. 1. *The Prisoner's Dilemma game. The payoff to player A is shown with illustrative numerical values. The game is defined by* $T > R > P > S$ *and* $R > (S + T)/2$.

(the temptation to defect). By assumption, $T > R$, so that it pays to defect if the other player cooperates. On the other hand, if the other player defects, there is a choice between cooperation which yields S (the sucker's payoff) or defection which yields P (the punishment for mutual defection). By assumption $P > S$, so it pays to defect if the other player defects. Thus, no matter what the other player does, it pays to defect. But, if both defect, both get P rather than the larger value of R that they both could have gotten had both cooperated. Hence the dilemma[17].

With two individuals destined never to meet again, the only strategy that can be called a solution to the game is to defect always despite the seemingly paradoxical outcome that both do worse than they could have had they cooperated.

Apart from being the solution in game theory, defection is also the solution in biological evolution[18]. It is the outcome of inevitable evolutionary trends through mutation and natural selection: if the payoffs are in terms of fitness, and the interactions between pairs of individuals are random and not repeated, then any population with a mixture of heritable strategies evolves to a state where all individuals are defectors. Moreover, no single differing mutant strategy can do better than others when the population is using this strategy. In these respects the strategy of defection is stable.

This concept of stability is essential to the discussion of what follows and it is useful to state it more formally. A strategy is evolutionarily stable if a population of individuals using that strategy cannot be invaded by a rare mutant adopting a different strategy[11]. In the case of the Prisoner's Dilemma played only once, no strategy can invade the strategy of pure defection. This is because no other strategy can do better with the defecting individuals than the P achieved by the defecting players who interact with each other. So in the single-shot Prisoner's Dilemma, to defect always is an evolutionarily stable strategy.

In many biological settings, the same two individuals may meet more than once. If an individual can recognize a previous interactant and remember some aspects of the prior outcomes, then the strategic situation becomes an iterated Prisoner's Dilemma with a much richer set of possibilities. A strategy would take the form of a decision rule which determined the probability of cooperation or defection as a function of the history of the interaction so far. But if there is a known number of interactions between a pair of individuals, to defect always is still evolutionarily stable and is still the only strategy which is. The reason is that defection of the last interaction would be optimal for both sides, and consequently so would defection on the next-to-last interaction, and so on back to the first interaction.

Our model is based on the more realistic assumption that the number of interactions is not fixed in advance. Instead, there is some probability, w, that after the current interaction the same two individuals will meet again. Factors that affect the magnitude of this probability of meeting again include the average lifespan, relative mobility, and health of the individuals. For any value of w, the strategy of unconditional defection (ALL D) is evolutionarily stable; if everyone is using this strategy, no mutant strategy can invade the population. But other strategies may be evolutionarily stable as well. In fact, when w is sufficiently great, there is no single best strategy regardless of the behavior of the others in the population[19]. Just because there is no single best strategy, it does not follow that analysis is hopeless. On the contrary, we demonstrate not only the stability of a given strategy, but also its robustness and initial viability.

Before turning to the development of the theory, let us consider the range of biological reality that is encompassed by the game theoretic approach. To start with, an organism does not need a brain to employ a strategy. Bacteria, for example, have a basic capacity to play games in that (i) bacteria are highly responsive to selected aspects of their environment, especially their chemical environment; (ii) this implies that they can respond differentially to what other organisms around them are doing; (iii) these conditional strategies of behavior can certainly be inherited; and (iv) the behavior of a bacterium can affect the fitness of other organisms around it, just as the behavior of other organisms can affect the fitness of a bacterium.

While the strategies can easily include differential responsiveness to recent changes in the environment or to cumulative averages over time, in other ways their range of responsiveness is limited. Bacteria cannot 'remember' or 'interpret' a complex past sequence of changes, and they probably cannot distinguish alternative origins of adverse or beneficial changes. Some bacteria, for example, produce their own antibiotics, bacteriocins; those are harmless to bacteria of the producing strain, but destructive to others. A bacterium might easily have production of its own bacteriocin dependent on the perceived presence of like hostile products in its environment, but it could not aim the toxin produced toward an offending initiator. From existing evidence, so far from an individual level, discrimination seems to be by species rather even than variety. For example, a *Rhizobium* strain may occur in nodules which it causes on the roots of many species of leguminous plants, but it may fix nitrogen for the benefit of the plant in only a few of these species[20]. Thus, in many legumes the *Rhizobium* seems to be a pure parasite. In the light of theory to follow, it would be interesting to know whether these parasitized legumes are perhaps less beneficial to free living *Rhizobium* in the surrounding soil than are those in which the full symbiosis is established. But the main point of concern here is that such discrimination by a *Rhizobium* seems not to be known even at the level of varieties within a species.

As one moves up the evolutionary ladder in neural complexity, game-playing behavior becomes richer. The intelligence of primates, including humans, allows a number of relevant improvements: a more complex memory, more complex processing of information to determine the next action as a function of the interaction so far, a better estimate of the probability of future interaction with the same individual, and a better ability to distinguish between different individuals. The discrimination of others may be among the most important of abilities because it allows one to handle interactions with many individuals without having to treat them all the same, thus making possible the rewarding of cooperation from one individual and the punishing of defection from another.

The model of the iterated Prisoner's Dilemma is much less restricted than it may at first appear. Not only can it apply to interactions between two bacteria or interactions between two primates, but it can also apply to the interactions between a colony of bacteria and, say, a primate serving as a host. There is no assumption of commensurability of payoffs between the two sides. Provided that the payoffs to each side satisfy the inequalities that define the Prisoner's Dilemma (Fig. 1), the results of the analysis will be applicable.

The model does assume that the choices are made simultaneously and with discrete time intervals. For most analytic purposes, this is equivalent to a continuous interaction over time, with the time period of the model corresponding to the minimum time between a change in behavior by one side and a response by the other. And while the model treats the choices as simultaneous, it would make little difference if they were treated as sequential[21].

Turning to the development of the theory, the evolution of cooperation can be conceptualized in terms of three separate questions:

(1) *Robustness*. What type of strategy can thrive in a variegated environment composed of others using a wide variety of more or less sophisticated strategies?

(2) *Stability*. Under what conditions can such a strategy, once fully established, resist invasion by mutant strategies?

(3) *Initial viability*. Even if a strategy is robust and stable, how can it ever get a foothold in an environment which is predominantly noncooperative?

ROBUSTNESS

To see what type of strategy can thrive in a variegated environment of more or less sophisticated strategies, one of us (R.A.) conducted a computer tournament for the Prisoner's Dilemma. The strategies were submitted by game theorists in economics, sociology, political science, and mathematics[22]. The rules implied the payoff matrix shown in Fig. 1 and a game length of 200 moves. The 14 entries and a totally random strategy were paired with each other in a round robin tournament. Some of the strategies were quite intricate. An example is one which on each move models the behavior of the other player as a Markov process, and then uses Bayesian inference to select what seems the best choice for the long run. However, the result of the tournament was that the highest average score was attained by the simplest of all strategies submitted: TIT FOR TAT. This strategy is simply one of cooperating on the first move and then doing whatever the other player did on the preceding move. Thus TIT FOR TAT is a strategy of cooperation based on reciprocity.

The results of the first round were then circulated and entries for a second round were solicited. This time there were 62 entries from six countries[23]. Most of the contestants were computer hobbyists, but there were also professors of evolutionary biology, physics, and computer science, as well as the five disciplines represented in the first round. TIT FOR TAT was again submitted by the winner of the first round, Professor Anatol Rapoport of the Institute for Advanced Study (Vienna). It won again. An analysis of the 3 million choices which were made in the second round identified the impressive robustness for TIT FOR TAT as dependent on three features: it was never the first to defect, it was provocable into retaliation by a defection of the other, and it was forgiving after just one act of retaliation[24].

The robustness of TIT FOR TAT was also manifest in an ecological analysis of a whole series of future tournaments. The ecological approach takes as given the varieties which are present and investigates how they do over time when interacting with each other. This analysis was based on what would happen if each of the strategies in the second round were submitted to a hypothetical next round in proportion to its success in the previous round. The process was then repeated to generate the time path of the distribution of strategies. The results showed that, as the less successful rules were displaced, TIT FOR TAT continued to do well with the rules which initially scored near the top. In the long run, TIT FOR TAT displaced all the other rules and went to fixation[24]. This provides further evidence that TIT FOR TAT's cooperation based on reciprocity is a robust strategy that can thrive in a variegated environment.

STABILITY

Once a strategy has gone to fixation, the question of evolutionary stability deals with whether it can resist invasion by a mutant strategy. In fact, we will now show that once TIT FOR TAT is established, it can resist invasion by any possible mutant strategy provided that the individuals who interact have a sufficiently large probability, w, of meeting again. The proof is described in the next two paragraphs.

As a first step in the proof we note that since TIT FOR TAT 'remembers' only one move back, one C by the other player in any round is sufficient to reset the situation as it was at the beginning of the game. Likewise, one D sets the situation to what it was at the second round after a D was played in the first. Since there is a fixed chance, w, of the interaction not ending at any given move, a strategy cannot be maximal in playing with TIT FOR TAT unless it does the same thing both at the first occurrence of a given state and at each resetting to that state. Thus, if a rule is maximal and begins with C, the second round has the same state as the first, and thus a maximal rule will continue with C and hence always cooperate with TIT FOR TAT. But such a rule will not do better than TIT FOR TAT does with another TIT FOR TAT, and hence it cannot invade. If, on the other hand, a rule begins with D, then this first D induces a switch in the state of TIT FOR TAT and there are two possibilities for continuation that could be maximal. If D follows the first D, then this being maximal at the start implies that it is everywhere maximal to follow D with D, making the strategy equivalent to ALL D. If C follows the initial D, the game is then reset as for the first move; so it must be maximal to repeat the sequence of

DC indefinitely. These points show that the task of searching a seemingly infinite array of rules of behavior for one potentially capable of invading TIT FOR TAT is really easier than it seemed: if neither ALL D nor alternation of D and C can invade TIT FOR TAT, then no strategy can.

To see when these strategies can invade, we note that the probability that the nth interaction actually occurs is w^{n-1}. Therefore, the expression for the total payoff is easily found by applying the weights $1, w, w^2 \ldots$ to the payoff sequence and summing the resultant series. When TIT FOR TAT plays another TIT FOR TAT, it gets a payoff of R each move for a total of $R + wR + w^2R \ldots$, which is $R/(1 - w)$. ALL D playing with TIT FOR TAT gets T on the first move and P thereafter, so it cannot invade TIT FOR TAT if

$$R/(1 - w) \geq T + wP/(1 - w)$$

Similarly when alternation of D and C plays TIT FOR TAT, it gets a payoff of

$$T = wS + w^2T + s^3 S \ldots$$
$$= (T + wS)/(1 - w^2)$$

Alternation of D and C thus cannot invade TIT FOR TAT if

$$R/(1 - w) \geq (T + wS)/(1 - w^2)$$

Hence, with reference to the magnitude of w, we find that neither of these two strategies (and hence no strategy at all) can invade TIT FOR TAT if and only if both

$$w \geq (T - R)/(T - P) \text{ and}$$
$$w \geq (T - R)/(R - S) \tag{1}$$

This demonstrates that TIT FOR TAT is evolutionarily stable if and only if the interactions between the individuals have a sufficiently large probability of continuing[19].

INITIAL VIABILITY

TIT FOR TAT is not the only strategy that can be evolutionarily stable. In fact, ALL D is evolutionarily stable no matter what is the probability of interaction continuing. This raises the problem of how an evolutionary trend to cooperative behavior could ever have started in the first place.

Genetic kinship theory suggests a plausible escape from the equilibrium of ALL D. Close relatedness of interactants permits true altruism— sacrifice of fitness by one individual for the benefit of another. True altruism can evolve when the conditions of cost, benefit, and relatedness yield net gains for the altruism-causing genes that are resident in the related individuals[25]. Not defecting in a single-move Prisoner's Dilemma is altruism of a kind (the individual is foregoing proceeds that might have been taken) and so can evolve if the two interactants are sufficiently related[18]. In effect, recalculation of the payoff matrix in such a way that an individual has a part interest in the partner's gain (that is, reckoning payoffs in terms of inclusive fitness) can often eliminate the inequalities $T > R$ and $P > S$, in which case cooperation becomes unconditionally favored[18,26]. Thus it is possible to imagine that the benefits of cooperation in Prisoner's Dilemma-like situations can begin to be harvested by groups of closely related individuals. Obviously, as regards pairs, a parent and its offspring or a pair of siblings would be especially promising, and in fact many examples of cooperation or restraint of selfishness in such pairs are known.

Once the genes for cooperation exist, selection will promote strategies that base cooperative behavior on cues in the environment[4]. Such factors as promiscuous fatherhood[27] and events at ill-defined group margins will always lead to uncertain relatedness among potential interactants. The recognition of any improved correlates of relatedness and use of these cues to determine cooperative behavior will always permit advance in inclusive fitness[4]. When a cooperative choice has been made, one cue to relatedness is simply the fact of reciprocation of the cooperation. Thus modifiers for more selfish behavior about a negative response from the other are advantageous whenever the degree of relatedness is low or in doubt. As such, conditionality is acquired, and cooperation can spread into circumstances of less and less relatedness. Finally, when the probability of two individuals meeting each other again is sufficiently high, cooperation based on reciprocity can thrive and be evolutionarily stable in a population with no relatedness at all.

A case of cooperation that fits this scenario, at least on first evidence, has been discovered in the spawning relationships in a sea bass[28]. The fish, which are hermaphroditic, form pairs and roughly may be said to take turns at being the high investment partner (laying eggs) and low investment partner (providing sperm to fertilize eggs). Up to ten spawnings occur in a day and only a few eggs are provided each time. Pairs tend to break up if sex roles are not divided evenly. The system appears to allow the evolution of much economy in the size of testes, but Fischer[28] has suggested that the testis condition may have evolved when the species was more sparse and inclined to inbreed. Inbreeding would imply relatedness in the pairs and this initially may have transferred the system to attractance of tit-for-tat cooperation—that is, to cooperation unneedful of relatedness.

Another mechanism that can get cooperation started when virtually everyone is using ALL D is clustering. Suppose that a small group of individuals is using a strategy such as TIT FOR TAT and that a certain proportion, p, of the interactions of members of this cluster are with other members of the cluster. Then the average score attained by the members of the cluster in playing the TIT FOR TAT strategy is

$$p[R/(1 - w)] + (1 - p)[S + wP/(1 - w)]$$

If the members of the cluster provide a negligible proportion of the interactions for the other individuals, then the score attained by those using ALL D is still $P/(1 - w)$. When p and w are large enough, a cluster of

TIT FOR TAT individuals can then become initially viable in an environment composed overwhelmingly of ALL D[19].

Clustering is often associated with kinship, and the two mechanisms can reinforce each other in promoting the initial viability of reciprocal cooperation. However, it is possible for clustering to be effective without kinship[3].

We have seen that TIT FOR TAT can intrude in a cluster on a population of ALL D, even though ALL D is evolutionarily stable. This is possible because a cluster of TIT FOR TAT's gives each member a nontrivial probability of meeting another individual who will reciprocate the cooperation. While this suggests a mechanism for the initiation of cooperation, it also raises the question about whether the reverse could happen once a strategy like TIT FOR TAT became established itself. Actually, there is an interesting asymmetry here. Let us define a nice strategy as one, such as TIT FOR TAT, which will never be the first to defect. Obviously, when two nice strategies interact, they both receive R each move, which is the highest average score an individual can get when interacting with another individual using the same strategy. Therefore, if a strategy is nice and is evolutionarily stable, it cannot be intruded upon by a cluster. This is because the score achieved by the strategy that comes in a cluster is a weighted average of how it does with others of its kind and with the predominant strategy. Each of these components is less than or equal to the score achieved by the predominant, nice, evolutionarily stable strategy, and therefore the strategy arriving in a cluster cannot intrude on the nice, evolutionarily stable strategy[19]. This means that when w is large enough to make TIT FOR TAT an evolutionarily stable strategy it can resist intrusion by any cluster of any other strategy. The gear wheels of social evolution have a ratchet.

The chronological story that emerges from this analysis is the following. ALL D is the primeval state and is evolutionarily stable. This means that it can resist the invasion of any strategy that has virtually all of its interactions with ALL D. But cooperation based on reciprocity can gain a foothold through two different mechanisms. First, there can be kinship between mutant strategies, giving the genes of the mutants some stake in each other's success, thereby altering the effective payoff matrix of the interaction when viewed from the perspective of the gene rather than the individual. A second mechanism to overcome total defection is for the mutant strategies to arrive in a cluster so that they provide a nontrivial proportion of the interactions each has, even if they are so few as to provide a negligible proportion of the interactions which the ALL D individuals have. Then the tournament approach demonstrates that once a variety of strategies is present, TIT FOR TAT is an extremely robust one. It does well in a wide range of circumstances and gradually displaces all other strategies in a simulation of a great variety of more or less sophisticated decision rules. And if the probability that interaction between two individuals will continue is great enough, then TIT FOR TAT is itself evolutionarily stable. Moreover, its stability is especially secure because it can resist the intrusion of whole clusters of mutant strategies. Thus cooperation based on reciprocity can get started in a predominantly noncooperative world, can thrive in a variegated environment, and can defend itself once fully established.

APPLICATIONS

A variety of specific biological applications of our approach follows from two of the requirements for the evolution of cooperation. The basic idea is that an individual must not be able to get away with defecting without the other individuals being able to retaliate effectively[29]. The response requires that the defecting individual not be lost in any anonymous sea of others. Higher organisms avoid this problem by their well-developed ability to recognize many different individuals of their species, but lower organisms must rely on mechanisms that drastically limit the number of different individuals or colonies with which they can interact effectively. The other important requirement to make retaliation effective is that the probability, w, of the same two individuals' meeting again must be sufficiently high.

When an organism is not able to recognize the individual with which it had a prior interaction, a substitute mechanism is to make sure that all of one's interactions are with the same interactant. This can be done by maintaining continuous contact with the other. This method is applied in most interspecies mutualism, whether a hermit crab and his sea-anemone partner, a cicada and the varied microorganismic colonies housed in its body, or a tree and its mycorrhizal fungi.

The ability of such partners to respond specifically to defection is not known but seems possible. A host insect that carries symbionts often carries several kinds (for example, yeasts and bacteria). Differences in the roles of these are almost wholly obscure[30]. Perhaps roles are actually the same, and being host to more than one increases the security of retaliation against a particular exploitative colony. Where host and colony are not permanently paired, a method for immediate drastic retaliation is sometimes apparent instead. This is so with fig wasps. By nature of their remarkable role in pollination, female fig wasps serve the fig tree as a motile aerial male gamete. Through the extreme protogyny and simultaneity in flowering, fig wasps cannot remain with a single tree. It turns out in many cases that if a fig wasp entering a young fig does not pollinate enough flowers for seeds and instead lays eggs in almost all, the tree cuts off the developing fig at an early stage. All progeny of the wasp then perish.

Another mechanism to avoid the need for recognition is to guarantee the uniqueness of the pairing of interactants by employing a fixed place of meeting. Consider, for example, cleaner mutualisms in which a small fish or a crustacean removes and eats ectoparasites from the body (or even from the inside of the mouth) of a larger fish which is its potential predator. These aquatic cleaner mutualisms occur in coastal and reef situations where animals live in fixed home ranges or territories[4,5]. They seem to be unknown in the free-mixing circumstances of the open sea.

Other mutualisms are also characteristic of situations where continued association is likely, and normally they involve quasi-permanent pairing of individuals or of endogamous or asexual stocks, or of individuals with such stocks[7,31]. Conversely, conditions of free-mixing and transitory pairing conditions where recognition is impossible are much more likely to result in exploitation—parasitism, disease, and the like. Thus, whereas ant colonies participate in many symbioses and are sometimes largely dependent on them, honeybee colonies, which are much less permanent in place of abode, have no known symbionts but many parasites[32]. The small freshwater animal *Chlorohydra viridissima* has a permanent stable association with green algae that are always naturally found in its tissues and are very difficult to remove. In this species the alga is transmitted to new generations by way of the egg. *Hydra vulgaris* and *H. attenuata* also associate with algae but do not have egg transmission. In these species it is said that[33] 'infection is preceded by enfeeblement of the animals and is accompanied by pathological symptoms indicating a definite parasitism by the plant'. Again, it is seen that impermanence of association tends to destabilize symbiosis.

In species with a limited ability to discriminate between other members of the same species, reciprocal cooperation can be stable with the aid of a mechanism that reduces the amount of discrimination necessary. Philopatry in general and territoriality in particular can serve this purpose. The phrase stable territories means that there are two quite different kinds of interaction: those in neighboring territories where the probability of interaction is high, and strangers whose probability of future interaction is low. In the case of male territorial birds, songs are used to allow neighbors to recognize each other. Consistent with our theory, such male territorial birds show much more aggressive reactions when the song of an unfamiliar male rather than a neighbor is reproduced nearby[34].

Reciprocal cooperation can be stable with a larger range of individuals if discrimination can cover a wide variety of others with less reliance on supplementary cues such as location. In humans this ability is well developed, and is largely based on the recognition of faces. The extent to which this function has become specialized is revealed by a brain disorder called prosopagnosia. A normal person can name someone from facial features alone, even if the features have changed substantially over the years. People with prosopagnosia are not able to make this association, but have few other neurological symptoms other than a loss of some part of the visual field. The lesions responsible for prosopagnosia occur in an identifiable part of the brain: the underside of both occipital lobes, extending forward to the inner surface of the temporal lobes. This localization of cause, and specificity of effect, indicates that the recognition of individual faces has been an important enough task for a significant portion of the brain's resources to be devoted to it[35].

Just as the ability to recognize the other interactant is invaluable in extending the range of stable cooperation, the ability to monitor cues for the likelihood of continued interaction is helpful as an indication of when reciprocal cooperation is or is not stable. In particular, when the value of w falls below the threshold for stability given in condition (1), it will no longer pay to reciprocate the other's cooperation. Illness in one partner leading to reduced viability would be one detectable sign of declining w. Both animals in a partnership would then be expected to become less cooperative. Aging of a partner would be very like disease in this respect, resulting in an incentive to defect so as to take a one-time gain when the probability of future interaction becomes small enough.

These mechanisms could operate even at the microbial level. Any symbiont that still has a transmission 'horizontally' (that is, infective) as well as vertically (that is, transovarial, or more rarely through sperm, or both) would be expected to shift from mutualism to parasitism when the probability of continued interaction with the host lessened. In the more parasitic phase it could exploit the host more severely by producing more infective propagules. This phase would be expected when the host is severely injured, contracted some other wholly parasitic infection that threatened death, or when it manifested signs of age. In fact, bacteria that are normal and seemingly harmless or even beneficial in the gut can be found contributing to sepsis in the body when the gut is perforated (implying a severe wound)[36]. And normal inhabitants of the body surface (like *Candida albicans*) can become invasive and dangerous in either sick or elderly persons.

It is possible also that this argument has some bearing on the etiology of cancer, insofar as it turns out to be due to viruses potentially latent in the genome[37]. Cancers do tend to have their onset at ages when the chances of vertical transmission are rapidly declining[38]. One oncogenic virus, that of Burkitt's lymphoma, does not have vertical transmission but may have alternatives of slow or fast production of infectious propagules. The slow form appears as a chronic mononucleosis, the fast as an acute mononucleosis or as a lymphoma[39]. The point of interest is that, as some evidence suggests, lymphoma can be triggered by the host's contracting malaria. The lymphoma grows extremely fast and so can probably compete with malaria for transmission (possibly by mosquitoes) before death results. Considering other cases of simultaneous infection by two or more species of pathogen, or by two strains of the same one, our theory may have relevance more generally to whether a disease will follow a slow, joint-optimal exploitation course ('chronic' for the host) or a rapid severe exploitation ('acute' for the host). With a single infection the slow course would be expected. With double infection, crash exploitation might, as dictated by implied payoff functions, begin immediately, or have onset later at an appropriate stage of senescence[40].

Our model (with symmetry of the two parties) could also be tentatively applied to the increase with maternal age of chromosomal nondisjunction during ovum formation (oogenesis)[41]. This effect leads to various conditions of severely handicapped offspring, Down's syndrome (caused by an extra copy of chromosome 21) being the most familiar example. It depends almost entirely on failure of the normal separation of the paired chromosomes in the mother, and this suggests the possible connection with our story. Cell divisions of oogenesis, but not usually of spermatogenesis, are characteristically unsymmetrical, with rejection (as a socalled polar body) of chromosomes that go to the unlucky pole of the cell.

It seems possible that, while homologous chromosomes generally stand to gain by steadily cooperating in a diploid organism, the situation in oogenesis is a Prisoner's Dilemma: a chromosome which can be 'first to defect' can get itself into the egg nucleus rather than the polar body. We may hypothesize that such an action triggers similar attempts by the homolog in subsequent meioses, and when both members of a homologous pair try it at once, an extra chromosome in the offspring could be the occasional result. The fitness of the bearers of extra chromosomes is generally extremely low, but a chromosome which lets itself be sent to the polar body makes a fitness contribution of zero. Thus $P > S$ holds. For the model to work, an incident of 'defection' in one developing egg would have to be perceptible by others still waiting. That this would occur is pure speculation, as is the feasibility of self-promoting behavior by chromosomes during a gametic cell division. But the effects do not seem inconceivable: a bacterium, after all, with its single chromosome, can do complex conditional things. Given such effects, our model would explain the much greater incidence of abnormal chromosome increase in eggs (and not sperm) with parental age.

Conclusion

Darwin's emphasis on individual advantage has been formalized in terms of game theory. This establishes conditions under which cooperation based on reciprocity can evolve.

1. Williams, G. C. *Adaptations and Natural Selection* (Princeton Univ. Press, Princeton, 1966); Hamilton, W. D. in *Biosocial Anthropology*, Fox, R. Ed. (Malaby, London, 1975), p. 133.
2. For the best recent case for effective selection at group levels and for altruism based on genetic correction of non-kin interactants see Wilson, D. S. *Natural Selection of Populations and Communities* (Benjamin/Cummings, Menlo Park, Calif., 1979).
3. Hamilton, W. D. *J. Theoret. Biol.* 7, 1 (1964).
4. Trivers, R. *Q. Rev. Biol.* 46, 35 (1971).
5. For additions to the theory of biological cooperation see Chase, I. D. [*Am. Nat.* 115, 827 (1980)], Fagen, R. M. [*ibid.*, p. 858 (1980)], and Boorman, S. A. and Levitt, P. R. [*The Genetics of Altruism* (Academic Press, New York, 1980)].
6. Dawkins, R. *The Selfish Gene* (Oxford Univ. Press, Oxford, 1976).
7. Hamilton, W. D. *Annu. Rev. Ecol. Syst.* 3, 193 (1972).
8. Janzen, D. H. *Evolution* 20, 249 (1966).
9. Wiebes, J. T. *Gard. Bull. (Singapore)* 29, 207 (1976); Janzen, D. H. *Annu. Rev. Ecol. Syst.* 10, 31 (1979).
10. Caullery, M. *Parasitism and Symbiosis* (Sidgwick and Jackson, London, 1952). This gives examples of antagonism in orchid-fungus and lichen symbioses. For the example of wasp-ant symbiosis, see (7).
11. Maynard Smith, J. & Price, G. R. *Nature (London)* 246, 15 (1973); Maynard Smith, J. and Parker, G. A. *Anim. Behav.* 24, 159 (1976); Parker, G. A. *Nature (London)* 274, 849 (1978).
12. Elster, J. *Ulysses and the Sirens* (Cambridge Univ. Press, London, 1979).
13. Emlen, S. T. in *Behavioural Ecology: An Evolutionary Approach*, Krebs, J. & Davies, N. Eds. (Blackwell, Oxford, 1978), p. 245; Stacey, P. B. *Behav. Ecol. Sociobiol.* 6, 53 (1979).
14. Harcourt, A. H. *Z. Tierpsychol.* 48, 401 (1978); Packer, C. *Anim. Behav.* 27, 1 (1979); Wrangham, R. W. *Soc. Sci. Info.* 18, 335 (1979).
15. Ligon, J. D. & Ligon, S. H. *Nature (London)* 276, 496 (1978).
16. Rapoport, A. & Chammah, A. M. *Prisoner's Dilemma* (Univ. of Michigan Press, Ann Arbor, 1965). There are many other patterns of interaction which allow gains for cooperation. See for example the model of intraspecific combat in Maynard Smith, J. & Price, G. R. in (11).
17. The condition that $R > (S + T)/2$ is also part of the definition to rule out the possibility that alternating exploitation could be better for both than mutual cooperation.
18. Hamilton, W. D. in *Man and Beast: Comparative Social Behavior* (Smithsonian Press, Wash-

ington, 1971), p. 57, Fagan, R. M. [in (5)] shows some conditions for single encounters where defection is not the solution.

19. For a formal proof, see Axelrod, R. *Am. Political Sci. Rev.*, in press. For related results on the potential stability of cooperative behavior see Luce, R. D. and Raiffa, H. *Games and Decisions* (Wiley, New York, 1957), p. 102; Taylor, M. *Anarchy and Cooperation* (Wiley, New York, 1976); Kurz, M. in *Economic Progress, Private Values and Public Policy*, Balassa, B. and Nelson, R. Eds. (North-Holland, Amsterdam, 1977), p. 177.

20. Alexander, M. *Microbiol Ecology* (Wiley, New York, 1971).

21. In either case, cooperation on a tit-for-tat basis is evolutionarily stable if and only if w is sufficiently high. In the case of sequential moves, suppose there is a fixed chance, p, that a given interactant of the pair will be the next one to need help. The critical value of w can be shown to be the minimum of the two side's value of $A/p(A + B)$ where A is the cost of giving assistance, and B is the benefit of assistance when received. See also Thompson, P. R. *Soc. Sci. Info.* **19**, 341 (1980).

22. Axelrod, R. *J. Conflict Resolution* **24**, 3 (1980).

23. In the second round, the length of the games was uncertain, with an expected probability of 200 moves. This was achieved by setting the probability that a given move would not be the last at $w = .99654$. As in the first round, each pair was matched in five games (24).

24. Axelrod, R. *J. Conflict Resolution* **24**, 379 (1980).

25. Fisher, R. A. *The Genetical Theory of Natural Selection* (Oxford Univ. Press, Oxford, 1930); Haldane, J. B. S. *Nature (London) New Biol.* **18**, 34 (1955); Hamilton, W. D. *Am. Nat.* **97**, 354 (1963).

26. Wade, M. J. & Breden, F. *Behav. Ecol. Sociobiol*, in press.

27. Alexander, R. D. *Annu. Rev. Ecol. Syst.* **5**, 325 (1974).

28. Fischer, E. *Anim. Behav.* **28**, 620 (1980); Leigh, Jr., E. G. *Proc. Natl. Acad. Sci. U.S.A.* **74**, 4542 (1977).

29. For economic theory on this point see Akerlof, G. Q. *J. Econ.* **84**, 488 (1970); Darby, M. R. and Karni, E. *J. Law Econ.* **16**, 67 (1973); Williamson, O. E. *Markets and Hierarchies* (Free Press, New York, 1975).

30. Buchner, P. *Endosymbiosis of Animals with Plant Microorganisms* (Interscience, New York, 1965).

31. Hamilton, W. D. in *Diversity of Insect Faunas*, Mound, L. A. and Waloff, N. Eds. (Blackwell, Oxford, 1978).

32. Wilson, E. O. *The Insect Societies* (Bellknap, Cambridge, Mass., 1971); Treisman, M. *Anim. Behav.* **28**, 311 (1980).

33. Yonge, C.M. [*Nature (London)* **134**, 12 (1979)] gives other examples of invertebrates with unicelluar algae.

34. Wilson, E. O. *Sociobiology* (Harvard Univ. Press, Cambridge, Mass., 1975), p. 273.

35. Geschwind, N. *Sci. Am.* **241**, (No. 3), 180 (1979).

36. Savage, D. C. in *Microbial Ecology of the Gut*. Clarke, R. T. J. and Bauchop, T. Eds. (Academic Press, New York, 1977), p. 300.

37. Manning, J. T. *J. Theoret. Biol.* **55**, 397 (1975); Orlove, M. J. *ibid.* **65**, 605 (1977).

38. Hamilton, W. D. *ibid.* **12**, 12 (1966).

39. Henle, W., Henle, G., Lenette, E. T. *Sci. Am.* **241** (No. 1) 48 (1979).

40. See also Eshel, I. *Theoret. Pop. Biol.* **11**, 410 (1977) for a related possible implication of multiclonal infection.

41. Stern, C. *Principles of Human Genetics* (Freeman, San Francisco, 1973).

42. For helpful suggestions we thank Robert Boyd, Michael Cohen, and David Sloan Wilson.

Robert Axelrod is a professor of political science and research scientist at the Institute of Public Policy Studies, University of Michigan and William D. Hamilton is a professor of evolutionary biology in the Museum of Zoology and the Division of Biological Sciences, University of Michigan.

This article first appeared in *Science* Vol. **211**, pp. 1390–1396; March 27, 1981. It is reproduced with permission. Copyright 1981 by the American Association for the Advancement of Science.

Mate choice increases a component of offspring fitness in fruit flies

LINDA PARTRIDGE

THE authors of two influential accounts of the genetic consequences of mate choice consider that animals cannot produce fitter offspring by mating with a fitter than average individual[1,2]. Their main reason for this is that in a population at genetic equilibrium any genetic variation which affects fitness should not be heritable[3]. Other accounts have suggested ways in which mate choice might improve offspring fitness[4]. Data are now presented which show that one component of offspring fitness can be increased by mate choice in *Drosophila melanogaster*. The component of fitness measured was intraspecific competitive success during the part of the life history from first instar larva to adult fly.

The flies used were an outbred stock collected in Dahomey in 1970 and maintained in approximately constant conditions in a population cage at the Department of Genetics, Edinburgh University. Two different experimental comparisons were made. In the first (experiment 1), the offspring of inseminated adult females which had mated in the population cage where many males were present were compared with offspring of females collected from the cage as virgins and mated with a randomly chosen male from the cage. The comparison was made by taking first instar larvae from eggs laid by the 'choice' and the 'no choice' females and competing each set of larvae against an equal number of larvae of Dahomey stock made homozygous for a fourth chromosome bearing the recessive mutant *sparkling*.

Larvae were allowed to compete in vials which contained the same food medium as that used in the population cage, and the number of wild-type and *sparkling* flies emerging from each vial was recorded. The point of this experimental procedure was to ensure that the wild-type parent flies were exactly matched for nutrition and numbers of anaesthetisations, and that the larvae were competing under circumstances similar to those encountered by larvae in the population cage, both with respect to the nature of the food and the genotype of the competitors. The experiment was done twice, and the results, shown in Table 1, indicate that in both replicates the percentage of wild-type flies emerging was slightly, but significantly, higher when the parents of those flies had chosen their mates.

Table 1 The competitive success of offspring of flies which had chosen (choice) or not chosen (no choice) their mates

| | | % Wild type | | No. of |
		Mean	Variance	vials
First replicate	Choice	51.1	7.1	23
	No choice	48.9	8.7	18
	$t = 2.45$	$P < 0.02$		
Second replicate	Choice	49.8	2.2	21
	No choice	48.1	5.8	14
	$t = 2.45$	$P < 0.02$		

EXPERIMENT 1. *The experimental details were as follows. Adult male and inseminated female flies were collected from the population cage using a pooter. After anaesthetisation with carbon dioxide, 100 males and 100 females were taken and placed individually in shell vials with food. Eclosing virgin females (100) were then collected from bottles in the population cage and placed individually in the 100 vials containing single males. After 2 days the 'no choice' females were separated from the males, and females of both groups were then set up in batches of 10 and allowed to lay eggs on 10% agar medium with a small amount of yeast present. At the same time, other batches of 10 females were set up using the Dahomey sparkling stock. All females were allowed to lay for 24 h before being moved on to fresh agar, and after a further 16 h first instar larvae were collected from the surface of the original agar using a paintbrush. The larvae were then placed in two groups of vials. Vials in the first group each contained 200 larvae from the 'choice' females and 200 sparkling larvae; vials in the second group each contained 200 larvae from the 'no choice' females together with 200 sparkling larvae. The vials were set up in this way over a period of 10 days, repeatedly taking offspring from the same females which used stored sperm during this period. It was therefore not known how many larvae each female contributed to each vial. The 400 larvae in each vial were allowed to compete and pupate, and the number of wild-type and sparkling flies emerging from each vial was recorded. Competitive success was measured as the percentage of wild-type (as opposed to sparkling) flies emerging from each vial. Significance levels are on a two-tailed t-test.*

Because the 'choice' flies in experiment 1 were collected at an older age than the 'no choice' females, experiment 2 was carried out to control for any effect of maternal age. First instar larvae were collected from parents of the same age who had either chosen or not chosen their mates, and the larvae were again competed against *sparkling* larvae. The results are shown in Table 2. Again, the percentage of wild-type flies emerging was larger if the parents of those flies had chosen their mates.

The results show that matings between randomly chosen pairs of flies produce offspring with lower survival between larvae and adult than matings where choice can occur. The results do not show whether the overall fitness of the offspring is affected, because some other component of fitness may be negatively correlated with the one measured.

The results could be produced by various genetic mechanisms. First, fitness may indeed be heritable, possibly as a consequence of mutational load[5]. If this were the case, the results could be produced in two ways. (1)

Table 2 The competitive success of offspring of females of the same age which had chosen (choice) or not chosen (no choice) their mates

| | % Wild type | | No. of |
	Mean	Variance	vials
Choice	50.45	2.0	28
No choice	49.44	2.0	32
$t = 2.72$	$P < 0.01$		

EXPERIMENT 2. *Adult flies were collected from the population cage using a pooter; after anaesthetisation with carbon dioxide the females were discarded, 100 males were placed individually in shell vials with food and 300 were placed in a single bottle with food. Eclosing females were then collected from bottles in the cage, and 100 were placed in the bottle containing 300 males and 100 individually in the 100 vials containing single males. The flies were left for 2 days, after which laying pots and competition vials were set up as in experiment 1. Competitive success was measured as the percentage of wild-type flies emerging in each vial. Probability level is on a two-tailed t-test.*

Fitter flies may be better at detecting or obtaining access to mates. Members of the other sex could then mate with a fly that was successful in some sort of competition with members of its own sex. (2) Flies may be able to detect heritable fitness in members of the other sex. Fitter flies could then be actively chosen as mates. Both possibilities would suggest that all flies should prefer the same sorts of mates. Alternatively, flies with high levels of heterozygosity may have high fitness. In this case, flies might produce fitter offspring by mating with individuals genetically unlike themselves, so that their offspring will have higher levels of heterozygosity. This last possibility would imply that genetically dissimilar flies would show different patterns of mate choice.

I thank Drs N. P. Ashmole, A. Ewing and L. Nunney and Professor A. Robertson for helpful advice.

1. Maynard Smith, J. *The Evolution of Sex* (Cambridge University Press, 1978).
2. Williams, G. C. *Sex and Evolution* (Princeton University Press, 1975).
3. Falconer, D. S. *Introduction to Quantitative Genetics* (Longman, London, 1960).
4. Trivers, R. L. in *Sexual Selection and the Descent of Man* (ed. Campbell, B.) 136–179 (Aldine-Atherton, Chicago, 1972).
5. Simmons, M. J. & Crow, J. F. *A. Rev. Genet.* **11**, 49 (1977).

Linda Partridge is in the Department of Zoology, University of Edinburgh.

This article first appeared in *Nature* Vol. **283**, pp. 290–291; 1980.

Testis weight, body weight and breeding system in primates

A. H. HARCOURT, P. H. HARVEY, S. G. LARSON
AND R. V. SHORT

It has long been known that primate species differ greatly in the weight of their testes relative to body weight[1]. Recently it has been suggested that among the three species of Pongidae (the great apes), the disparity in testes weights is associated with their different breeding systems[2-4]. Male gorillas and orangutans copulate infrequently, and when a female comes into oestrus she normally mates with only one male. However, in the chimpanzee, several males mate frequently with the oestrous females, so that each male has to deposit enough sperm to compete with the presence of sperm from other males. For the chimpanzee, therefore, we hypothesize that selection will favour the male that can deposit the largest number of sperm; thus the volume of spermatogenic tissue and hence testis size is far greater in the chimpanzee than in the gorilla or orangutan. If this is correct, it implies that primates in which more than one male mates with each oestrous female should have larger testes relative to their body weight than those with single-male breeding systems. We have tested this prediction across a wide range of primates, and the results support the hypothesis. The relative size of testes may, therefore, provide a valuable clue to the breeding system of a primate species.

The male gorilla (*Gorilla gorilla*) and orangutan (*Pongo pygmaeus*), heaviest of the primates, have breeding systems that involve one male monopolizing mating with a number of females[5-7], and have testes that together weigh 30 g and 35 g respectively. The lighter chimpanzee (*Pan troglodytes*) male, by contrast, has a breeding system in which ~75% of copulations and at least 25% of conceptions occur during periods of promiscuous mating when several males copulate with each oestrous female[8] (multi-male breeding system); its combined testis weight is ~120 g. It has been suggested that the marked differences in testes weights among the three species are related to their breeding systems[2-4]. In the single-male breeding system of the gorilla and orangutan, each male need ejaculate only enough sperm to ensure fertilization, whereas in the multi-male system of the chimpanzee, each male also has to inseminate sufficient to compete with the presence of sperm from other males. Thus, selection in the chimpanzee for high sperm production may explain why it has larger testes than the gorilla or orangutan. High sperm storage could also be selected for, but we have too few data to test this idea.

An obvious test of the hypothesis relating size of testes to breeding system is to determine whether a similar association is found in other primates. This has been done for baboons[9], but the sample size was small and allometric effects were not taken into account. We therefore examined a wide range of primate species representing 18 genera from six families, from a 320-g marmoset (*Callithrix*) to the 170-kg gorilla. Data from 33 species taken from the literature and some unpublished results are shown in Table 1. No prosimian species were included in the statistical analysis because of lack of reliable information on their mating systems. Only data from fully mature individuals were included, and the body weights are those of the individuals whose testes were weighed. Where more than one specimen per species was available, the testes weight of the individual of median body weight was used, except for *Saimiri sciureus* and *Presbytis entellus* where the original sources required use of, respectively, the mean and the midpoint between extreme values. Note that throughout we considered breeding, not social, systems. The distinction is relevant for species such as the gorilla[5], that can occur in multi-male groups but have single-male breeding systems.

Testes weight increases with body weight, even for species with the same breeding system (Fig. 1). We therefore corrected for the effect of body weight by examining deviations from the line of best fit. The major axis line was used because some error was incorporated into measures of both body weight and testes weight, and because the data were logarithmically transformed[10]. The taxonomic level analysed was the genus. Generic values were calculated from the means of the logarithmically transformed values for species, but congenerics were kept separate if they had different breeding systems. The species level was not used in statistical analysis because closely related species within various genera have very similar values and cannot be treated as independent points. In addition, when the species level was analysed, we found significant heterogeneity in deviations due to family membership (taxonomic effect) within the monogamous group. Thus, analysis at the genus level allowed us to examine the relationship between testes weight and breeding system, with body weight and taxonomic effects removed.

As no significant heterogeneity in slope among breeding systems was revealed by maximum likelihood analysis[10] ($\chi^2 = 3.10$), a common slope of 0.66 was fitted through all the points (Fig. 1). Furthermore, within breeding systems, there was no heterogeneity of deviations from the line among taxa (monogamous, $F_{2,1} = 2.57$; single male, $F_{2,4} = 2.43$; multi male, $F_{2,5} = 3.33$; one-way analysis of variance). Among breeding systems, however, as predicted, we found significant heterogeneity of deviations from the common slope ($F_{2,16} = 10.29$; $P < 0.01$); whereas monogamous and single-male genera do not differ from each other in their deviations about the common line ($t_5 = 0.62$; Student's t-test), both show the predicted significant differences from the multi-male genera ($t_6 = 2.45$, $P < 0.05$ and $t_{15} = 4.76$, $P < 0.01$, respectively). As the deviations were measured

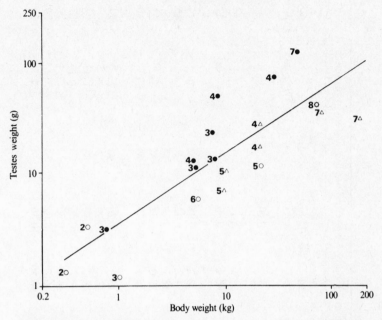

FIG. 1. *Log combined testes weight (g) versus log body weight (kg) for different primate genera.* ●, *Multi-male breeding system;* ○, *monogamous;* △, *single-male,* ⊙, *Homo. Numbers represent the families shown in Table 1. (4) and (5) were combined for statistical analysis.*

perpendicular to the body weight axis, the results show that multi-male genera have significantly heavier testes in relation to body weights than do the single-male or monogamous genera.

The positive correlation between testes weight and body weight, having a common slope of < 1.0, is true of many organ to body weight relationships among mammals[11]. Also, in this study a significant correlation between the two variables was found in three of the eight species for which records from > 10 specimens were available (median $N = 17.5$; $P < 0.05$, Spearman rank correlation coefficients). Among species, such a correlation is expected because the testes are endocrine glands whose output and hence volume has to increase with body size if threshold concentrations of hormones are to be maintained. In addition, larger species require more sperm to counteract the dilution effect of the larger volume of the male and female reproductive tracts[4].

The hypothesis that competition in sperm numbers is important in the association of testes size with breeding system requires that the multi-male primates, as well as having larger testes, have a greater volume of seminiferous tubules, rather than interstitial tissue, and greater sperm production capabilities than the one-male or monogamous species. This seems to be the case; *Pan*, *Papio* (baboon) and *Macaca* (macaque), which all

Table 1 Combined testes weight (excluding epididymis), body weight and mating system of primates

	N	Body weight (kg)	Testes weight (g)	Ratio (%)	Mating system	Refs Weights	Refs Mating system
(1) Lorisidae							
Loris tardigradus (w)	114	0.27	1.8	0.66	?	14	
(2) Callitrichidae							
Callithrix jacchus (c)	15	0.32	1.3	0.41	P	*	15
Sanguinus oedipus (w)	56	0.52	3.4	0.65	P	1, 16	17
(3) Cebidae							
Saimiri sciureus (w)	40	0.78	3.2	0.41	M	18	19
Aotus trivirgatus (w)	1	1.02	1.2	0.12	P	1	15
Lagothrix lagothricha (c)	1	5.22	11.2	0.2	M	1	15, 20
Alouatta palliata (w)	8	7.26	23.0	0.32	M	21	22
Ateles geoffroyi (w)	1	7.94	13.4	0.17	M	1	15
(4) Cercopithecinae							
Cercopithecus aethiops (w)	6	4.95	13.0	0.26	M	23	24
Cercocebus atys (c)	1	8.68	25.1	0.29	?	25	
Macaca fascicularis (w, c)	20	4.42	35.2	0.80	M	1, †	26
Macaca radiata (w, c)	2	8.65	48.2	0.56	M	1, ‡	27
Macaca mulatta (c)	26	9.2	46.2	0.50	M	1, 12, ‡	28
Macaca nemestrina (w)	1	9.98	66.7	0.67	M	1	29
Macaca arctoides (c)	20	10.51	48.15	0.46	M	‡	29

Papio hamadryas (w)	6	20.17	27.1	0.13	S	23	30
Papio cynocephalus (c)	21	24.32	52.0	0.21	M	‡	31
Papio anubis (w)	4	26.40	93.5	0.35	M	23	32
Papio ursinus (?)	1	31.75	72.0	0.23	M	33	34
Papio papio (w)	1	31.98	88.9	0.28	M	1	35
Theropithecus gelada (w)	1	20.40	17.1	0.08	S	23	36
(5) Colobinae							
Presbytis rubicunda (w)	12	6.23	3.4	0.05	?	1	37
Presbytis cristata (w)	12	6.58	6.2	0.09	S	1	38
Presbytis obscura (w)	14	7.45	4.8	0.06	S	§	38
Presbytis entellus (w)	6	17.0	11.1	0.06	S	39, 40	41
Colobus polykmos (= *guereza*) (w)	3	10.25	10.7	0.10	S	19	42
Nasalis larvatus (w)	8	20.64	11.8	0.06	S	1	43, [1]
(6) Hylobatidae							
Hylobatus moloch (w)	7	5.44	6.1	0.11	P	1	44
Hylobates lar (c)	1	5.5	5.5	0.10	P	45	44
(7) Pongidae							
Pan troglodytes (c)	3	44.34	118.8	0.27	M	1	8
Pongo pygmaeus (w)	2	74.64	35.3	0.05	S	1	6, 7
Gorilla gorilla (w)	2	169.0	29.6	0.02	S	46, 47	5
(8) Hominidae							
Homo sapiens	4	65.65	40.5	0.06	S, P	1, 48	49

(w), Specimens caught from the wild; (c), captive specimens. P, monogamous (pair); M, multi-male; S, single-male breeding system. Unpublished data from *R.V.S.; †P. Squires; ‡S.G.L.; §G. J. Burton and Earl Cranbrook; [1]M. Cutler and S. M. Jeffrey.

have multi-male breeding systems, have ratios of tubules to connective tissue ranging from 2.2 to 2.8:1, whereas *Homo*, *Hylobates* (gibbon) and *Presbytis* (langur), which are all non-multi-male species, have ratios of only 0.9 to 1.3:1 (ref. 1). In addition, the sperm production rate of the multi-male rhesus macaque (*Macaca mulatta*), with its high volume of tubules, is 23×10^6 sperm per g of testis per day, whereas in man it is only 4.4×10^6 sperm per g per day[12,13].

Differences in breeding system are clearly not the only reason for variation in testes weight among the primates. Seasonality of breeding, for example, could be an important factor: species with a short breeding season, which have a high level of copulatory activity for a period, probably need larger testes. Certainly the multi-male genus *Macaca*, a predominantly seasonal breeder, shows the largest deviation from the common line. Nevertheless, there is clearly a relationship between testes weight and breeding system, despite exceptions such as *Saguinus oedipus* (tamarin); thus it seems reasonable to predict that similar trends will be found in other mammalian orders.

We thank Drs J. Clevedon Brown, T. H. Clutton-Brock, R. D. Martin and B. P. Setchell for valuable advice and criticism; and the following for unpublished data: Dr G. J. Burton, Earl Cranbrook and the University of Malaya, P. Squires, M. Cutler, S. M. Jeffrey and the Huntingdon Research Centre.

1. Schultz, A. H. *Anat. Rec.* **72**, 387–394 (1938).
2. Short, R. V. in *Reproduction and Evolution* (eds Calaby, J. H. & Tyndale-Biscoe, C. H.) (Australian Academy of Science, Canberra, 1977).
3. Short, R. V. *Adv. Study Behav.* **9**, 131–158 (1979).
4. Short, R. V. in *Reproductive Biology of the Great Apes* (ed. Graham, C. E.) (Academic, New York, 1981).
5. Harcourt, A. H. in *Reproductive Biology of the Great Apes* (ed. Graham, C. E.) (Academic, New York, 1981).
6. Rijksen, H. D. *A Fieldstudy on Sumatran Orang Utans* (Veenman & Zonen, Wageningen, 1978).
7. Gladikas, B. M. F. in *Reproductive Biology of the Great Apes* (ed. Graham, C. E.) (Academic, New York, 1981).
8. Tutin, C. E. G. *J. Reprod. Fert.* Suppl. **28**, 43–57 (1980).
9. Collins, A. *New Scient.* **78**, 12–14 (1978).
10. Harvey, P. H. & Mace, G. M. in *Current Problems in Sociobiology* (ed. King's College Research Centre Sociobiology Project) (Cambridge University Press, in the press).
11. Brody, S. *Bioenergetics and Growth* (Reinhold, New York, 1945).
12. Amann, R. P., Johnson, L., Thompson, D. L. & Pickett, B. W. *Biol. Reprod.* **15**, 586–592 (1976).
13. Amann, R. P. & Howards, S. S. *J. Urol.* **124**, 211–215 (1980).
14. Ramakrishna, P. A. & Prasad, M. R N. *Folia primatol.* **5**, 176–189 (1967).
15. Moynihan, M. *The New World Primates* (Princeton University Press, New Jersey, 1976).
16. Dawson, G. A. & Dukelow, W. R. *J. med Primatol.* **5**, 266–275 (1976).
17. Moynihan, M. *Smithson. Contr. Zool.* **28**, 1–77 (1970).
18. Middleton, C. C. & Rosal, J. *Lab. Anim.* **22**, 583–586 (1972).
19. Baldwin, J. D. *Folia primatol.* **9**, 281–314 (1968).
20. Kavanagh, M. & Dresdale, L. *Primates* **16**, 285–294 (1975).
21. Hrdlicka, A. *Am. J. phys. Anthrop.* **8**, 201–211 (1925).
22. Chivers, D. J. *Folia primatol.* **10**, 48–102 (1969).
23. Kinsky, M. *Anat. Anz.* **108**, 65–82 (1960).
24. Struhsaker, T. T. *Behaviour* **29**, 83–121 (1967).
25. Kennard, M. A. & Willner, M. D. *Endocrinology* **28**, 977–984 (1941).
26. Angst, W. in *Primate Behavior, Developments in Field and Laboratory Research* Vol. 4 (ed. Rosenblum, L. A.) (Academic, New York, 1975).
27. Simonds, P. E. in *Primate Behavior. Field Studies of Monkeys and Apes* (ed. DeVore, I.) (Holt, Rinehart & Winston, New York, 1965).
28. Southwick, C. H., Beg, M. A. & Siddiqi, M. R. in *Primate Behavior. Field Studies of Monkeys and Apes* (ed. DeVore, I.) (Holt, Rinehart & Winston, New York, 1965).
29. Roonwal, M. L. & Mohnot, S. M. *Primates of South Asia* (Harvard University Press, Massachusetts, 1977).
30. Kummer, H. *Social Organisation of Hamadryas Baboons* (University of Chicago Press, 1968).

31. Hausfater, G. *Contr. Primatol.* **7**, 1–150 (1975).
32. Hall, K. R. L. & DeVore, I. in *Primate Behavior. Field Studies of Monkeys and Apes* (ed. DeVore, I.) (Holt, Rinehart & Winston, New York, 1965).
33. Hill, W. C. O. *Primates. Comparative Anatomy and Taxonomy 8, Cynopithecinae* (Edinburgh University Press, 1970).
34. Hall, K. R. L. *Proc. zool. Soc. Lond.* **139**, 283–327 (1962).
35. Dunbar, R. I. M. & Nathan, M. F. *Folia primatol.* **17**, 321–334 (1972).
36. Dunbar, R. I. M. & Dunbar, P. *Contr. Primatol.* **6** (1975).
37. Bernstein, I. S. *Behaviour* **32**, 1–16 (1968).
38. Curtin, S. H. thesis, Univ. Calif., Berkeley (1976).
39. David, G. F. X. & Ramaswami, L. S. *J. Morph.* **135**, 99–130 (1971).
40. Braz, I., Shandilya, L. N. & Ramaswami, L. S. *Andrologia* **8**, 290–296 (1976).
41. Hrdy, S. B. *The Langurs of Abu* (Harvard University Press, Massachusetts, 1977).
42. Oates, J. F. *Z. Tierpsychol.* **45**, 1–160 (1977).
43. Kawabe, M. & Mano, T. *Primates* **13**, 213–227 (1972).
44. Chivers, D. J. in *Primate Conservation* (eds Prince Rainier III & Bourne, G. H.) (Academic, New York, 1977).
45. Kennard, M. A. & Willner, M. D. *Endocrinology* **28**, 967–976 (1941).
46. Wislocki, G. B. *J. Mammal.* **23**, 281–287 (1942).
47. Hall-Craggs, E. C. B. *Proc. zool. Soc. Lond.* **139**, 511–514 (1962).
48. Benoit, J. *C. r. Séanc Soc. Biol.* **87**, 1387–1390 (1922).
49. Daly, M. & Wilson, M. *Sex, Evolution and Behavior* (Duxbury, Massachusetts, 1978).

A. H. Harcourt is in the Department of Applied Biology, University of Cambridge; P. H. Harvey in the School of Biological Sciences, University of Sussex; S. G. Larson in the Department of Anthropology, University of Wisconsin, Madison and R. V. Short in the MRC Reproductive Biology Unit, Edinburgh.

This article was first published in *Nature* Vol. **293**, pp. 55–57; 1981.

Outward signs of breeding

ROBERT D. MARTIN AND ROBERT M. MAY

ANIMAL behaviour, both within and between species, tends to exhibit much greater variability than does morphology or physiology. This makes it difficult to reconstruct phylogenies from behaviour patterns. The problems are particularly acute for mammals, where social behaviour can be very variable in its expression, depending on local ecological conditions and on prevailing demographic factors. Discussion of the origins of mammalian social systems therefore benefits greatly from any linkage to morphological or physiological features, which provide anchor points that are relatively stable in evolutionary terms.

A good example is provided by Harvey and colleagues' analysis of sexual dimorphism of body size among primates[1]. Studies of living primates (and, incidentally, of other mammals) show that sexual dimorphism is absent in all species which exhibit habitual monogamy, whereas it may or may not be present in species with social systems where males have breeding access to more than one female (unimale or multimale polygyny). Accordingly, the apparent presence of sexual dimorphism in some of the earliest known (Oligocene) relatives of the Old World monkeys and apes from the Fayum deposits of Egypt suggests that they possessed polygynous mating systems of some kind[2].

Focusing on the Great Apes—orang-utan, gorilla, chimpanzee and man—Short has drawn together information about several morphological features that shed light on the relationships between sexual selection and behavioural ecology[3-5]. This example is especially interesting because the four species are genetically and biochemically very similar, yet phenotypically and behaviourally significantly different. The data assembled by Short are summarized schematically in the figure, which shows the sexual dimorphism in body size, the size and location of the testes, the relative size of the erect penis, and the relative development of the mammary glands and the perineum, for each of the four species.

As discussed by Short, these morphological characteristics correlate well with the breeding system observed for orang-utans, gorillas and chimpanzees (we will return to man below), and these in turn may be plausibly associated with the foraging behaviour of the females of the various species[6].

Orang-utans are large arboreal frugivores, and the density of fruiting trees or those with edible leaves and bark is not particularly high in the

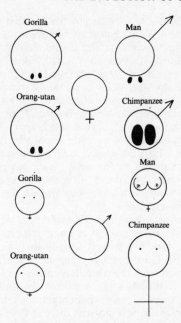

In the top half of the figure, the size of the circles shows the body size of a breeding male, relative to the size of a typical female (central circle), for each of the four species; it also shows the size and location of the testes and the relative size of the erect penis (arrow on circle). The lower half of the figure similarly shows the relative body sizes of the females compared to a typical male, along with schematic depiction of the relative development of the mammary glands and of the perineum before the first pregnancy (cross beneath circle). From Short[3, 4].

thick forests these creatures inhabit. Female orang-utans thus live in isolation and males occupy large 'core areas' which typically contain several females and from which they exclude other adult males. The mating system is thus basically a unimale polygynous one (effectively a harem group). There is consequently selection on the male for the physical size and strength to defend his core area, resulting in a pronounced sexual dimorphism where male orang-utans are about twice the size of females. On the other hand, copulation is a relatively rare event[7]; each of the few adult females within earshot of a given male is likely to come into heat only once or twice in 6 years or so (the mean birth interval is about 5–7 years). At such low copulation frequencies, there is no need for a high spermatogenic capacity—hence the relatively small testes. These circumstances, together with the lack of visibility in the forest canopy, put little selective pressure on sexual advertisement. Noting that the erect penis is about 4 cm long, Short[5] adds the engaging observation that: 'To what extent, if any, it is used in male–female display is uncertain, but its length certainly allows the animal to adopt a wide variety of copulatory positions, necessitated by the fact that copulation usually occurs when both animals are hanging from branches'.

Gorillas are predominantly terrestrial, and feed on rather low-quality herbage which is usually abundant but in fairly widely dispersed patches. The females are thus relatively less concerned about competition for food and travel in small groups. Males compete for exclusive ownership of such groups (which fragment on the death of the dominant male). This polygynous, unimale breeding system is basically similar to that for the orang-utan, even though the ecological details are quite different. Again there is strong sexual dimorphism, little sexual advertisement (the erect penis is about 3 cm long) and small testes. Indeed, a typical male gorilla has three to four females in his troop, each having roughly 4 year intervals between births, so that he will encounter an oestrous female on average but once a year[8]; relative to body weight, the gorilla testis is the smallest of any known primate.

Chimpanzees are terrestrial and arboreal omnivores, and live in relatively open habitats where vision is more important. However, the availability of their food is relatively low, so each female needs a large core area; thus, rather like the female orang-utan, the female chimpanzee tends to remain apart from the male. It is infeasible for a single male to defend a territory that embraces several females and the outcome is a system in which groups of males cooperate to defend a large area containing many females. In this system, there is little aggression among males within the group, which may explain the low degree of sexual dimorphism. Such social cooperation carries the consequence that males share the sexual favours of females that come into oestrus; chimpanzees have a promiscuous, multimale breeding system. Estimates[5] of the size of the community of females available to a male, coupled with estimates of the frequency of oestrus for a given female, suggest that intercourse is almost a daily occurrence for a male chimpanzee, in vivid contrast to the gorilla and the orang-utan where it is roughly an annual event. The consequent emphasis on testis size, and sexual advertisement in both male and female, are as depicted in the figure.

One unsatisfactory aspect of the above story is that the comparisons are among so few species. In the preceding article, Harcourt and co-workers present a very nice study of the relationship between size of testes and type of mating system for some 33 species of monkeys and apes. As with recent studies of sexual dimorphism, an effective comparison of testis size among so varied a group of primates can only be conducted after the scaling influence of body size itself has been taken into account[9]. This reflects a theme in much of developmental biology: mechanical constraints on the way living machines can be constructed will tend to give general 'allometric' laws governing the way particular characteristics scale with body weight; superimposed on these broad patterns will be variations arising from adaptation to particular ecological or behavioural circumstances. Thus, for mammalian reproduction in general, it has recently been demonstrated that many different aspects of fertility (litter size, litter frequency and so on) exhibit systematic dependence on environmental circumstances (tropical versus temperate; tree-dwelling versus burrowing) once the systematic effects of body weight are removed[10].

Harcourt *et al.*'s analysis confirms the suggestion by Short[3-5] that, once body size has been properly taken into account, primate species with multimale breeding systems consistently have larger testes than those with unimale breeding systems (monogamous pairs or harem groups). As explained more fully by Harcourt *et al.*, this is expected: a greater capacity for sperm production will have obvious advantages wherever males may encounter direct competition within their social group for breeding access to females. It therefore appears that the relative size of the testes is a very useful guide to the basic breeding system of any primate species, reflecting fundamental adaptation regardless of minor adjustments of social behaviour to suit local ecological conditions. Extension of these comparisons to mammals generally should yield further useful insights, and a more reliable base for interpretation.

As in any approach dealing with only a single parameter relative to body size, there are complications, the most obvious of which is seasonality. For any given breeding system, males of a species which exhibits mating for only a restricted part of the year might be expected to possess larger testes than males of species breeding throughout the year. In addition, testis size is known to vary over the year in seasonally breeding mammal species. Any such seasonal effects must evidently be taken into account for a really clear evaluation of the implications of relative testis size. For example, among callitrichids it is known that the cotton-top tamarin (*Saguinus oedipus*) exhibits a marked seasonal breeding peak, whereas the common marmoset (*Callithrix jacchus*) breeds throughout the year with no obvious seasonal pattern[11]. This might explain why cotton-top tamarins have relatively larger testes than common marmosets, despite the fact that both species are classified as monogamous. Similarly, Harcourt *et al.* point out that *Macaca* species, which generally show some kind of seasonal breeding peak, exhibit the largest relative testis size among all the primate species included in the comparison. There is also an apparent enigma in that the squirrel monkey (*Saimiri sciureus*) is known to exhibit very pronounced seasonality of breeding and is also classified as possessing a multimale breeding system, yet its testes (relative to body size) appear to be only moderately developed and are in fact both relatively and absolutely smaller than in the monogamous cotton-top tamarin. However, in this case it is possible that the data used by Harcourt and colleagues, which were derived from a laboratory colony, do not accurately reflect the natural situation. For this reason alone, it is unfortunate that no data on prosimian primates could be included in their analysis, since some of the more extreme forms of seasonal reproduction known among the primates are found in the Madagascar lemurs, with some species exhibiting mating for only a very few months each year[12]. It can be predicted that among lemurs, testis size (relative to body size) should be larger than in monkeys and apes generally, but that monogamous species such as the indri (*Indri indri*) should show relatively small testes compared with other lemur species, such as the ringtail (*Lemur catta*) and the sifaka (*Propithecus verreauxi*), which live in multimale breeding units.

Reliance on testis size as an indicator of breeding systems also begs the question of the timing of mating relative to ovulation. There is some evidence from baboons that dominant males mate with females closer to the time of ovulation than do subordinate males[13], and it is not at all clear whether increased semen volume or better timing of mating would offer the better chances of breeding success for a particular male in a multimale breeding group. For example, if male A were to mate with a given female 48 hours before ovulation, while male B were to mate only 24 hours before ovulation, it is not known, at present, which male's spermatozoon would fertilize the egg.

As noted by Harcourt et al., a male's success in competing with other males for mating with an oestrous female will not depend only on his sperm production capacity, as reflected by testis size. Increased sperm storage capacity will also enhance a male's chances of success in mating competition and one might therefore expect primate species with multimale breeding systems to exhibit a larger relative size of the epididymis than species with unimale breeding systems. Data are not yet available on epididymis weights in primates, but this is an obvious priority for future investigation.

What does all this tell us about ourselves? Man is unusual in that every mating system known for the order Primates as a whole can be found among contemporary human societies. Even if attention is restricted to those simpler existing societies that are uncontaminated by Western culture, the evidence remains equivocal: an analysis[14] of 185 such societies found 74 per cent to be basically polygynous, although economic considerations and a shortage of women meant that in practice about half adopted monogamy. Indeed, some argue that human mating systems do not have a biological basis, but rather the form adopted in any given society is dictated largely by socioeconomic factors. On the other hand, the popular literature on human behaviour shows well established support for the notion that there is a biological basis for monogamy in *Homo sapiens*, and elaborate scenarios have been constructed around this idea.

The analysis conducted by Harcourt et al. shows that human testis size certainly does not accord with the range of values to be expected for a multimale breeding system. This line of evidence, however, does not discriminate between monogamy (whether serial or lifelong) and polygyny (unimale harem groups); relatively small testis size is found for monogamous primates such as the gibbon, and also, as illustrated in the figure, for polygynous primates such as the gorilla and the orang-utan.

Two other lines of evidence suggest, however, that our hunter-gatherer ancestors were likely to have been polygynous rather than monogamous. First, modern *Homo sapiens* exhibits sexual dimorphism both in body size (albeit to a mild degree, with men about 20 per cent heavier than women at any given age) and in form[5]. This contrasts with the well established rule that sexual dimorphism is not found among monogamous mammals. Second, as has been discussed with specific reference to the red deer[15] (*Cervus elaphus*), it is to be expected that parents will allocate a greater proportion of their resources to sons rather than to daughters only in cases where 'reproductive success varies more widely among males than females'. For human gestation, it is known that more resources are devoted to a male fetus than to a female one, resulting in significantly higher birth-weights for male infants[16]. Since there is no cultural influence which might affect the developing fetus differentially according to its sex, this may be taken as indicating some biological basis for a mating system in which human males would exhibit more variation in reproductive success than would females; that is, polygyny or promiscuity rather than monogamy. Setting aside the trends in degree of sexual dimorphism and relative weight of testes, the other patterns in the figure are mainly concerned with the conspicuousness of the genitalia (size of erect penis, whether or not the testes are pendulous and so on). These patterns have not yet been compared systematically among primate species, and the evidence they offer about mating systems can be subjected to an indecisive variety of interpretations.

Note the special importance of the study by Harcourt *et al.* in these attempts to reconstruct the breeding systems of our human ancestors by broad comparisons among primates. Earlier work tends to rule out monogamy, but makes little distinction between multimale and unimale breeding groups; the current study of the relative size of testes now tends to exclude multimale systems. Monogamy may, of course, nonetheless be the optimal system in certain modern societies. Much work on birds, mammals and other creatures indicates that monogamy often emerges when large investments in the offspring necessitate cooperation between the parents. It is possible that the increasing dependence of the human infant, associated with progressively increasing brain size and cultural complexity, has favoured culturally determined monogamous tendencies in various human societies, even though these have not led to the complete suppression of biological indicators of a polygynous ancestry.

In brief, it seems likely that our morphology is that of a species with unimale breeding groups but that these biological antecedents are today often overlain by extremely powerful socioeconomic determinants.

1. Harvey, P. H., Kavanagh, M. & Clutton-Brock, T. H. *J. Zool. Lond.* **186**, 475 (1978).
2. Martin, R. D. *Nature, News & Views* **287**, 273 (1980).
3. Short, R. V. *Adv. Study Behav.* **9**, 131 (1979).
4. Short, R. V. *J. Reprod. Fert. Suppl.* **28**, 3 (1980).
5. Short, R. V. in *Reproductive Biology of the Great Apes* (ed. Graham, C. E.) 319 (Academic, New York, 1981).
6. Wrangham, R. W. *Soc. Sci. Inf.* **18**, 335 (1976).
7. Galdikas, B. M. F. in *The Great Apes* (eds Hamburg, D. A. & McCown, E. R.) 195 (Benjamin, 1979).
8. Harcourt, A. H., Fossey, D., Stewart, K. J. & Watts, D. P. *J. Reprod. Fert. Suppl.* **28**, 59 (1980).
9. Martin, R. D. *Z. Morph. Anthrop.* **71**, 115 (1980).
10. May, R. M. & Rubenstein, D. I. in *Reproductive Biology*, Vol. 6, 2nd edn (eds Austin, M. & Short, R. V.) (Cambridge University Press, 1981).
11. Brand, H. M. *Lab. Anim.* **14**, 301 (1980).
12. Van Horn, R. N. *Prog. reprod. Biol.* **5**, 181–221 (1980).
13. Rowell, T. E. in *Social Communication among Primates* (ed. Altmann, S. A.) 15 (University of Chicago Press, 1967).
14. Ford, C. S. & Beach, F. A. *Patterns of Sexual Behaviour* (Eyre & Spottiswood, London, 1952).
15. Clutton-Brock, T. H., Albon, S. D. & Guiness, F. E. *Nature* **289**, 487 (1981).
16. Gibson, J. R. & McKeown, T. *Br. J. soc. med.* **6**, 152 (1952).

Robert D. Martin is Reader in Physical Anthropology at University College, London and Robert M. May is Professor of Biology at Princeton University.

This article first appeared in *Nature* Vol. **293**, pp. 8–9; 1981.